CRITICAL THINKING & LOGICAL REASONING WORKBOOK-2

GIFT OF LOGIC™ SERIES

Boost Your Thinking Skills

An Essential Resource for Everyone

Verbal Reasoning

Analytical Reasoning

Pictorial Reasoning

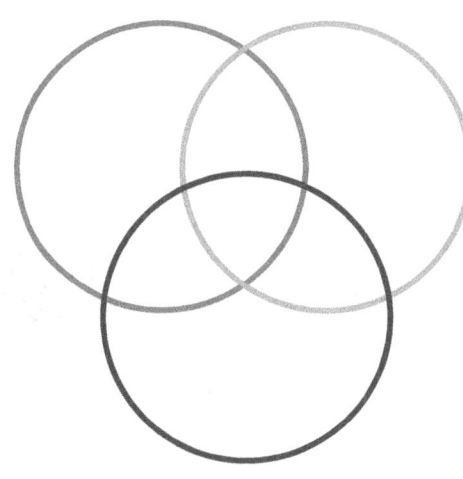

THIRD EDITION

| FOR GRADES K-2 | STUDENTS, TEACHERS, AND PARENTS |

Ranga Raghuram 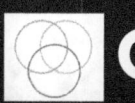 **GIFT OF LOGIC™**

Gift Of Logic, Inc

http://www.giftoflogic.com
sales@giftoflogic.com

Critical Thinking and Logical Reasoning Workbook-2
ISBN-13: 978-1494831110
ISBN-10: 1494831112

Third Edition
1-2014

Copyright © 2009 Gift Of Logic, Inc. All rights reserved. No part of this publication may be reproduced, stored in a retrieval system, transmitted in any form or by any means, electronic, mechanical, photocopying, recording or otherwise, without the written permission of the publisher.

License: This book is licensed for use by one person only. Use of this book in a group setting (classroom, workshop, etc) without the written permission of the publisher is prohibited. Unauthorized duplication is strictly prohibited by law. Contact the publisher at sales@giftoflogic.com for classroom/school/group licensing.

GIFT OF LOGIC™
CRITICAL THINKING & LOGICAL REASONING CURRICULUM
12 WORKBOOKS TO BOOST YOUR THINKING SKILLS

For Kindergarten, Grade 1, and Grade 2

Workbook# 0

Verbal Reasoning	Finding the truth, Inferencing, Analogies, Synonyms and Antonyms, Agree/Disagree
Analytic Reasoning	Memory drill, Decision making, Positioning, Sudoku
Pictorial Reasoning	Connect the dots, Mazes, Picture Sequence, Spot the difference, etc

Workbook# 1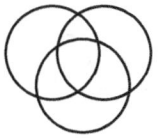

Verbal Reasoning	Finding the truth, Inferencing, Analogies, Synonyms and Antonyms, Agree/Disagree
Analytic Reasoning	Sorting, Positioning, Picking, Assorted problems, Numeric and Alphabetic Sudoku
Pictorial Reasoning	Picture Sequence, Spot the difference, Odd picture

Workbook# 2

Verbal Reasoning	Finding the truth, Classification, Direct and Inverse relationship, Inferencing, Analogies, Agree/Disagree
Analytic Reasoning	Sequencing, Scheduling, Strategy, Picking, etc
Pictorial Reasoning	Picture Analogy, Odd picture, Pattern matching, etc

For Grade 3, Grade 4, and Grade 5

Workbook# 3

Verbal Reasoning	Not, And, Or, If .. then, Conditional inferencing, Unconditional inferencing, Symbolic Logic
Analytic Reasoning	Lists, Sequencing, Grouping, Venn Diagrams, Graph logic, Number logic, Letter logic, Sudoku
Pictorial Reasoning	Picture sequence, Picture analogy, Odd picture, Picture difference, Pattern matching

Workbook# 4

Verbal Reasoning	Contradiction, Converse, Inverse, Contrapositive, Conditional inferencing, Symbolic Logic
Analytic Reasoning	Scheduling, Looping, FIFO, LIFO, Correlation, Venn Diagram, Graph logic, Number logic, Sudoku, etc
Pictorial Reasoning	Picture sequence, Picture analogy, Odd picture, Picture difference, Pattern matching

Workbook# 5

Verbal Reasoning	Biconditional, Categorical inferencing, Cause and Effect, Symbolic Logic, Agree/Disagree, Word and Sentence analogy
Analytic Reasoning	Correlation, Grouping, Venn Diagrams, Graph logic, Number logic, Letter logic, Sudoku, etc
Pictorial Reasoning	Picture sequence, Picture analogy, Odd picture, Picture difference, Pattern matching

********* Essential resource for everyone *********
*http://www.giftoflogic.com   *sales@giftoflogic.com

GIFT OF LOGIC™
CRITICAL THINKING & LOGICAL REASONING CURRICULUM
12 WORKBOOKS TO BOOST YOUR THINKING SKILLS

For Grades 6-12, College/University Students, Adults

Primer

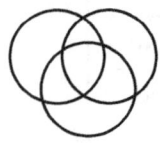

Prereq

Verbal Reasoning	Logical Operators, Conditional, Categorical and Causal reasoning, Validity, Fallacies, Symbolic Logic
Analytic Reasoning	Positioning, Grouping, Sudoku
Pictorial Reasoning	Pattern perception, Figure formation, Paper folding and cutting, Figure matrix, Rule detection

Workbook# 6

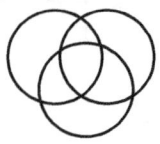

Verbal Reasoning	Arguments-Main point, Must be true, Cannot be true
Analytic Reasoning	Positioning, Grouping, Sudoku
Pictorial Reasoning	Pattern perception, Figure formation, Paper folding and cutting, Figure matrix, Rule detection

Workbook# 7

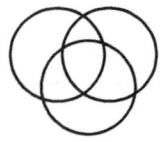

Verbal Reasoning	Arguments-Strengthening, Weakening
Analytic Reasoning	Positioning, Grouping, Sudoku
Pictorial Reasoning	Pattern perception, Figure formation, Paper folding and cutting, Figure matrix, Rule detection

Workbook# 8

Verbal Reasoning	Arguments - Controversy, Paradox
Analytic Reasoning	Positioning, Grouping, Sudoku
Pictorial Reasoning	Pattern perception, Figure formation, Paper folding and cutting, Figure matrix, Rule detection

Workbook# 9

Verbal Reasoning	Arguments- Assumptions, Reasoning strategy
Analytic Reasoning	Positioning, Grouping, Sudoku
Pictorial Reasoning	Pattern perception, Figure formation, Paper folding and cutting, Figure matrix, Rule detection

Workbook# 10

Verbal Reasoning	Arguments-Flawed reasoning, Analogous reasoning
Analytic Reasoning	Positioning, Grouping, Sudoku
Pictorial Reasoning	Pattern perception, Figure formation, Paper folding and cutting, Figure matrix, Rule detection

********* Essential resource for everyone *********
Get the GIFT OF LOGIC™ today !
*http://www.giftoflogic.com *sales@giftoflogic.com

Dear Reader:

Your decision to purchase this book is commendable. You now have in your hands, a comprehensive, easy-to-read book in Critical thinking and Logical reasoning that will introduce you to three different areas of thinking and reasoning - Verbal, Analytical and Pictorial. Solving problems in Verbal Reasoning is important to develop a critical mind. Solving problems in Analytic Reasoning is important to develop a flexible and resourceful mind. Solving problems in Pictorial Reasoning is important to develop a visually alert mind.

This book is presented in a workbook format to help you progress quickly. Parents and teachers are urged to complete the exercises ahead of the student and assist them whenever necessary with the help of detailed answers provided at the end of the book. This book can be used as a supplementary resource in the regular class room or it can be used during winter and summer vacations. College/University students, working professionals and retired individuals will also find the Gift Of Logic(tm) Series very useful in enhancing their problem solving abilities, confidence and general intellect.

Critical thinking and Logical reasoning must be practiced consistently to develop strong cognitive skills. After completing the exercises in this book, continue to read the other books in this series to get familiar with different types of Logical reasoning problems.

This workbook is one in a series of twelve workbooks. Please refer to the brochure before this page for a brief description of each workbook. Visit the website http://www.giftoflogic.com for more information.

Happy thinking and reasoning!

TABLE OF CONTENTS

Verbal Reasoning

Finding the truth..9

Direct or Inverse relationship..13

Classification..15

Word Analogy...17

Sentence Analogy...19

Agree or Disagree...21

Inferencing (Deductive Reasoning)
    Must be true...23
    Cannot be true...27

Analytic Reasoning

Inferencing (True or False) 30

Sequencing...32

Scheduling..43

Strategy problems...45

TABLE OF CONTENTS

Analytic Reasoning (continued)

Positioning problems..52

Picking problems..57

Sudoku
    Numeric Sudoku..59
    Alphabetic Sudoku...65

Pictorial Reasoning

Picture Sequence..72

Picture Difference..78

Picture Analogy..80

Odd Picture..83

Pattern Matching...87

Answers

Verbal...91
Analytic..105
Pictorial...144

Certificate of Completion

Name _____   Date _____

VERBAL REASONING

Name _____   Date _____

FINDING THE TRUTH

Finding the truth of statements is important for reasoning correctly.
Find the truth of the following statements and circle the correct answer.
Do some research if necessary.

1	I wake up before 7 AM on Tuesdays. A) True B) False
2	I have never woken up before 7 AM. A) True B) False
3	Except on Wednesdays, I play a game of Chess everyday. A) True B) False
4	Except Mars, all the other planets orbit around the Sun. A) True B) False
5	John can ride his bike on a lane exclusively reserved for cars. A) True B) False
6	A soccer game that will be shown exclusively in channel 19, can also be seen in channel 14. A) True B) False

Verbal Reasoning   Answers-91
© Gift Of Logic, Inc * Copying prohibited

Name _____   Date _____

FINDING THE TRUTH

Finding the truth of statements is important for reasoning correctly.
Find the truth of the following statements and circle the correct answer.
Do some research if necessary.

7	A turtle is amphibious. A) True B) False
8	The days between Sunday and Friday, inclusive, are the days of the week. A) True B) False
9	There are three fingers between the thumb finger and the little finger. A) True B) False
10	There are twelve months between January and December. A) True B) False
11	There are twelve months between January and December, inclusive. A) True B) False
12	Almost all birds can fly. A) True B) False
13	A majority of votes is required to win an election. A) True B) False

Verbal Reasoning   Answers-91
© Gift Of Logic, Inc * Copying prohibited

Name _____ Date _____

FINDING THE TRUTH

Finding the truth of statements is important for reasoning correctly.
Find the truth of the following statements and circle the correct answer.
Do some research if necessary.

14	Ten inches of rope is longer than five inches of iron rod. A) True B) False
15	Eight pounds of sugar is heavier than seven pounds of rice. A) True B) False
16	Ten gallons of water is more than nine gallons of water. A) True B) False
17	A big square is smaller than a small square. A) True B) False
18	Moon is closer to Earth than it is to Sun. A) True B) False
19	An item that costs ten dollars is less expensive than an item that costs seven dollars. A) True B) False

Verbal Reasoning   Answers-92
© Gift Of Logic, Inc * Copying prohibited

Name _____   Date _____

FINDING THE TRUTH

Finding the truth of statements is important for reasoning correctly.
Find the truth of the following statements and circle the correct answer.
Do some research if necessary.

20	If object A is inside object B, then object B must be inside object A. A) True B) False
21	Tiger and Lion are the same type of animals. A) True B) False
22	Our ears are attached to the head, and the head is attached to the body. A) True B) False
23	Earth is part of the solar system, and mountains are part of earth. A) True B) False
24	Muscles are not part of bones. A) True B) False
25	Snakes and lizards belong to the reptile family. A) True B) False

Verbal Reasoning   Answers-92
© Gift Of Logic, Inc * Copying prohibited

Name _____   Date _____

DIRECT OR INVERSE RELATIONSHIP

A direct relationship between two things exists when one thing increases when the other thing also increases or when one thing decreases when the other thing also decreases.

An inverse relationship between two things exists when one thing increases when the other thing decreases or when one thing decreases when the other thing increases.

In the following questions, two sets of items are given. Evaluate the relationship between them as Direct or Inverse..

1	The number of books read - The number of words learned A) Direct B) Inverse
2	Amount of time we are awake - Amount of time we are asleep A) Direct B) Inverse
3	Road accidents - Traffic delays A) Direct B) Inverse
4	Money spent - Money saved A) Direct B) Inverse
5	Time for playing - Time for studies A) Direct B) Inverse

Verbal Reasoning   Answers-94
© Gift Of Logic, Inc * Copying prohibited

Name _____   Date _____

DIRECT OR INVERSE RELATIONSHIP

Read the following statements and find if it describes a direct or an inverse relationship.

A direct relationship between two things exists when one thing increases when the other thing also increases or when one thing decreases when the other thing also decreases.

An inverse relationship between two things exists when one thing increases when the other thing decreases or when one thing decreases when the other thing increases.

6	Amount of rain - Scarcity of water A) Direct B) Inverse	
7	Number of babies - Population of earth. A) Direct B) Inverse	
8	Efficiency - Time A) Direct B) Inverse	
9	Number of pages in a book - Book reading time A) Direct B) Inverse	
10	Number of vehicles - Air pollution A) Direct B) Inverse	

Verbal Reasoning   Answers-94
© Gift Of Logic, Inc * Copying prohibited

Name _____   Date _____

CLASSIFICATION

Classification: Classification means to sort the items into groups so that all the items in a group have the same property. In the following questions, four items are given. Three of them can be assigned to a group whereas one item does not belong to that group and is the odd item. Find the odd item and circle it.

1	A) Eye	B) Nose	C) Teeth	D) Nail
2	A) Governor	B) President	C) Mayor	D) Actor
3	A) Car	B) Bike	C) Truck	D) Van
4	A) January	B) February	C) June	D) July
5	A) Boat	B) Ship	C) Submarine	D) Yacht
6	A) Air	B) Water	C) Milk	D) Oil
7	A) Mechanic	B) Electrician	C) Thief	D) Postman
8	A) Hut	B) Temple	C) Church	D) Lawn
9	A) Inches	B) Gram	C) Meter	D) Feet
10	A) Earth	B) Sun	C) Venus	D) Jupiter
11	A) North	B) Left	C) South	D) East

Verbal Reasoning   Answers-96

		CLASSIFICATION		
12	A) Stove	B) Microwave	C) Fridge	D) Bed
13	A) Swimming	B) Football	C) Hockey	D) Volleyball
14	A) Piano	B) Keyboard	C) Guitar	D) Flute
15	A) Rhino	B) Lion	C) Elephant	D) Giraffe
16	A) Year	B) Tomorrow	C) Month	D) Day
17	A) Italian	B) German	C) French	D) Canadian
18	A) English	B) Spanish	C) French	D) Algebra
19	A) Dog	B) Cat	C) Fox	D) Squirrel
20	A) Parrot	B) Ostrich	C) Eagle	D) Crow
21	A) River	B) Ocean	C) Lake	D) Puddle
22	A) Paper	B) Pen	C) Pencil	D) Magnet
23	A) Cover	B) Poem	C) Page	D) Binding
24	A) Actor	B) Princess	C) King	D) Brother
25	A) Cup	B) Pot	C) Plate	D) Bottle

Verbal Reasoning   Answers-96

© Gift Of Logic, Inc * Copying prohibited

Name _____   Date_____

WORD ANALOGY

The first two words separated by a colon (:) have a specific relationship. The third word has the same relationship with one of the words in the answer choices given. Circle the answer choice that will complete the analogy.

1	buy : get => sell : A) take B) give C) throw
2	train : rail => ship : A) water B) air C) road
3	affection : good => anger : A) nice B) bad C) helpful
4	flower : blossom => tree : A) break B) grow C) fruit
5	mouth : cough => nose : A) spit B) sneeze C) suck
6	rat : rodent => snake : A) reptile B) wild C) bug

Verbal Reasoning   Answers-97

© Gift Of Logic, Inc * Copying prohibited

Name _____   Date _____

WORD ANALOGY

The first two words separated by a colon (:) have a specific relationship. The third word has the same relationship with one of the words in the answer choices given. Circle the answer choice that will complete the analogy.

7	teeth : grind => hand : A) kick B) squeeze C) bite
8	radio : sound => television : A) picture B) art C) food
9	ocean : whale => pond : A) fish B) shark C) octopus
10	trash : burn: => valuables : A) throw B) keep C) hide
11	fruit : juice => corn : A) water B) oil C) chemical

Verbal Reasoning   Answers-97
© Gift Of Logic, Inc * Copying prohibited

Name _____   Date_____

SENTENCE ANALOGY

The following sentences contain an analogy. Spot the analogy and answer the questions.

1

Mark: I feel like a deflated balloon.

What is the analogy?

Mark feels
    A) strong
    B) weak

2

Sherry: I can run like a horse.

What is the analogy?

Sherry can run
    A) fast
    B) slow

---

3   Brian: Shelly sings like a cuckoo.

What is Brian comparing?

According to Brian, Shelly's voice is
    A) nice to hear
    B) hard to bear

Verbal Reasoning   Answers-98

© Gift Of Logic, Inc * Copying prohibited

Name _____   Date_____

SENTENCE ANALOGY

4

John held Mike's hand like a crocodile.

What is the comparison?

John held Mike's hand
    A) lightly.
    B) tightly.

5

Maggie: Susan won a gold medal in the singing competition.
Pam: This is music to my ears.

What is Pam comparing?

Pam says that the news of Susan winning a gold medal is
    A) bad news to her.
    B) good news to her.

6

The news spread like wildfire.

What is being compared?

Which one of the following is true?
    A) news spread fast.
    B) news spread slowly.

Name _____   Date _____

AGREE-DISAGREE

Read the statements of two people and find out whether they both agree or disagree with each other.

---

1 Jack: Apple is the tastiest fruit.
   Jill: An Orange tastes better than an apple.

Jack and Jill
   A) agree with each other that apple is the tastiest fruit.
   B) disagree with each other that apple is the tastiest fruit.

2

Mary: Everyone must read this book once.
Mark: Everyone must read this book more than once.

Mary and Mark
  A) agree on the number of times the book must be read.
  B) disagree on the number of times the book must be read.

---

3 Walid: We should not be tardy to school.
   Mohan: We should be punctual to school.

Walid and Mohan
   A) agree with each other.
   B) disagree with each other.

Verbal Reasoning   Answers-100
© Gift Of Logic, Inc * Copying prohibited

Name _____   Date _____

AGREE-DISAGREE

4

Mia: Every child must attend school to learn.
Lee: Every child must attend school to make friends.

Mia and Lee
  A) agree with each other on the purpose of going to school.
  B) disagree with each other on the purpose of going to school.

Mia and Lee
  A) agree with each other that every child must go to school.
  A) disagree with each other that every child must go to school.

5

Raja: We must sleep a lot to remain healthy.
Rashid: We must work a lot remain healthy.

Raja and Rashid.
  A) have the same opinion on how to remain healthy.
  B) have different opinions on how to remain healthy.

6

Sonia: We can use as much electricity as we want.
Maggie: We must be frugal while using electricity.

Sonia and Maggie
  A) agree on how to use electricity.
  B) disagree on how to use electricity.

Verbal Reasoning   Answers-100
© Gift Of Logic, Inc * Copying prohibited

Name _____   Date _____

| INFERENCE - must be true |

Inferencing means finding from a given set of facts, new information that is true, but not stated. In questions 1-10, some facts are given. Assuming that these facts are true, find out which of the given inferences must be true.

1 divide

A country is divided into several states. A state is divided into several cities.

Which one of the following must be true about a city?
   A) It belongs to a state and a country.
   B) It belongs to a state only.

2 every

Every car driver has a driver's license. A driver's license shows the picture, name, and address of the driver.

Which one of the following must be true?
   A) If a man has an address, he will have a driver's license.
   B) If a man has a driver's license, he will have an address.

3 except

Except the first grade and third grade, every other grade can go to the museum.

Which one of the following must be true?
   A) A first grade and a fourth grade student can go to the museum.
   B) A second grade and a fourth grade student can go to the museum.

Verbal Reasoning   Answers-102
© Gift Of Logic, Inc * Copying prohibited

Name _____   Date _____

| INFERENCE - must be true |

4 will, except

The recycling truck will come every day except on weekends.

Which one of the following must be true?
    A) The recycling truck will come on Saturdays.
    B) The recycling truck will not come on Sundays.

5 classification

Botany is a branch of Biology. Zoology is also a branch of Biology. Plants and trees are studied in Botany. Animals are studied in Zoology.

Which one of the following must be true?
    A) Lions are studied in Botany.
    B) Oak trees are studied in Botany.

6 belongs to

The Green Leaf party and the Blue Leaf party contested in the elections. Baker belongs to the Blue Leaf party. Gary belongs to the Green Leaf party. The Blue Leaf party won the election.

Which one of the following must be true?
    A) Gary was the winner.
    B) Baker was the winner.

Verbal Reasoning   Answers-102

Name _____   Date _____

| INFERENCE - must be true |

7 at least

Residents of Richland were told that gold can be found after digging the earth to a depth of two hundred feet. John, who resides in Richland, found gold while digging in his backyard.

If the above statements are true, which one of the following must be true?
   A) John dug at least two hundred feet.
   B) John dug less than two hundred feet.

8 increase/decrease

The population of a city grew for five years from 1990 and then it remained the same for another five years.

If the above statements are true, which one of the following must be true?
   A) More people lived in the city in 2000 than in 1995.
   B) More people lived in the city in 2000 than in 1990.

Name _____   Date _____

| INFERENCE - must be true |

9 maximum limit

The new Liberty High School can have a maximum of 200 boys and 150 girls. Within a few days after the admissions began, this limit was reached.

If the above statements are true, which one of the following must be true?
- A) 300 students will attend the school.
- B) More than 300 students will attend the school.

10 same/different

Physical education classes are held on Mondays and Fridays every week.

If the above statement is true, which one of the following must be true?
- A) Physical education classes are held on the same days every week.
- B) Physical education classes are held on different days every week.

Verbal Reasoning   Answers-103
© Gift Of Logic, Inc * Copying prohibited

Name _____   Date _____

INFERENCE - cannot be true

You should know how to spot an incorrect inference. In questions 11-14, some facts are given. Assuming that these facts are true, find out which of the inferences given cannot be true. Use your common sense to answer these questions.

11 timing

An emergency phone call was made to the fire department at 9 PM. A fire truck immediately rushed to the spot and put out the fire.

Which one of the following cannot be true?
    A) The fire was put out at 8 PM.
    B) The fire was put out after 9 PM.

---

12 same time

The Creek Park has five soccer fields. A soccer game is played by two teams.

Which one of the following cannot be true?
    A) Five games can be played at the same time.
    B) Five teams can play at the same time.

INFERENCE - cannot be true

13
first come first served

Jason was the first in line at the bookstore to buy a popular novel. Books are sold on a first come-first served basis.

Which one of the following cannot be true?
A) John, who was also in the line, was the first one to purchase the novel.
B) The novel that was sold to Jamie, who was also in the line, was not the first novel to be sold.

14
speed

Two cars, one red and one blue, started from the same spot at the same time and raced for one hundred miles. The speed of the red car was more than the speed of the blue car.

Which one of the following cannot be true?
A) The red car won the race.
B) The blue car won the race.

ANALYTICAL REASONING

Name _____   Date _____

INFERENCE - true or false

1 age

The following table shows the dress code for students five years to eight years old.

Blue pants	Green pants	Red pants
5-6 years	6-7 years	7-8 years

Ray is six and a half years old. So, he will wear red color pants.

The inference (conclusion) is
      A) True   B) False

---

2 height

A road goes inside a tunnel that is twelve feet high. So, a cement truck taller than twelve feet can travel in this road.

The inference (conclusion) is
      A) True   B) False

---

3 ends

A magnet has a north pole and a south pole. Similar poles of two magnets repel each other, but opposite poles of two magnets attract each other. So, when the south pole of a red magnet touches a blue magnet and stays attached to it, then the pole of the blue magnet that is in contact with the red magnet is its north pole.

The inference (conclusion) is
      A) True   B) False

| INFERENCE - true or false |

4 and

Jean is six feet tall, Joan is five feet tall, and Jane is four feet tall. Therefore, Jean is taller than Joan and Joan is shorter than Jane.

The inference (conclusion) is

        A) True   B) False

---

5 inside

Circle A and circle B are completely inside circle C. Therefore, circle A is completely inside circle B.

The inference (conclusion) is
        A) True   B) False

Name _____   Date _____

SEQUENCING - order

1

sequence -time/alphabet

* Write the days of the week in sequence starting from Sunday.

* Write the months of the year, in sequence, starting from the first month.

* Write all the objects in a table that you can see, in ascending alphabetic order.

* Write the name of all members of your family in descending order of their first names.

* Write the names of five cities in your state in descending order.

* Write all the places you visited yesterday in sequence from morning to evening.

Name _____ Date _____

SEQUENCING - time

2

Three movies will be screened at exactly three 3 hour intervals starting at 4 PM. Each movie is of 2 hours duration.

1) When will each of the three movies begin?

2) At what time will the second movie end?

3) At what time will the last movie end?

3

A theater complex has two theaters A and B. Three movies will be screened in theater A starting at 4 PM, two hours apart. Three movies will be screened in theater B starting at 5 PM, two hours apart.

1) At what time will the fourth movie in the complex be screened?

2) In which theater will the second movie in the complex be screened?

3) When and where will the last movie in the complex be screened?

Analytical Reasoning   Answers-108
© Gift Of Logic, Inc * Copying prohibited

Name _____   Date _____

SEQUENCING - simultaneous

4

A sequence of red, green and blue lights flash consecutively in Row# 1. Simultaneously, a sequence of yellow, blue and orange lights flash consecutively in row# 2 below. All the bulbs flash for the same duration.

1) At any particular time, how many lights flash?
   A) 2   B) 3   C) 6

2) When the blue light in Row#2 has finished flashing, how many lights would have finished flashing
   - in Row# 1?
   - in Row# 2?
   - in Total?

3) When the blue light in row#1 flashes, which one of the following lights in row#2 also will flash?

   A) Blue   B) Yellow   C) Orange

4) Before the orange light starts flashing, how many lights would have finished flashing in total?
   A) 3   B) 4   C) 5

Analytical Reasoning   Answers-109
© Gift Of Logic, Inc * Copying prohibited

Name —————————————— Date ——————————————

SEQUENCING - repeat, every other

5 repeat

A loud drum beat is represented by the symbol ! and a soft drum beat is represented by symbol ~. A sequence of one loud drum beat and two soft drum beats was repeated 5 times.

1) Represent the sequence using symbols.

2) The fifth drum beat was loud.
    A) True   B) False

3) Every even drum beat was soft and every odd drumbeat was loud.
    A) True   B) False

---

6 every other

Several ants marched up towards a sugar cube in the kitchen floor in sequence. Starting with second ant, every other ant is female.

If all the ants up to the third female ant in the sequence are separated from the rest of the ants, the number of ants that would be separated is

    A) 3   B) 5   C) 6

Name _____ Date _____

SEQUENCING - chart, before, after

7

The following chart shows the sequence of classes held Monday to Friday from 9 AM to Noon at a school.

	9 AM	10 AM	11 AM	Noon
Monday	Math	Science	Science	Reading
Tuesday	Reading	Reading	Math	Science
Wednesday	Math	Math	Reading	Science
Thursday	Science	Math	Science	Math
Friday	Reading	Science	Math	Reading

1) How many science classes start before noon on Monday?

2) How many math classes start after 11 AM?

3) How many science classes are held before Wednesday?

4) How many Reading classes start before 11 AM Thursday?

5) Write the sequence of classes held at noon.

Analytical Reasoning   Answers-111
© Gift Of Logic, Inc * Copying prohibited

SEQUENCING - before, after, first, last

8

The following is eye doctor Frank's schedule for today.

Time	Patient
9 AM	Laura
9:30 AM	Steve
10 AM	Roger
10:30 AM	Mary
11 AM- 2 PM	Break
2 PM	Mark
2:30 PM	Ron

1) How many patients will the doctor see today?

2) The first appointment for Dr. Frank is before 9 AM.
     A) True   B) False

3) Mary will be seen by Dr. Frank before Mark is seen.
     A) True   B) False

4) Laura and Roger will see the doctor before Steve.
     A) True   B) False

5) Laura and Steve will see the doctor after Mary and Ron.
     A) True   B) False

Analytical Reasoning   Answers-111

Name _____ Date _____

SEQUENCING - first, next, before, after, between

9 Seven cars tagged as C1,C2,C3,C4,C5,C6, and C7 were lined up one behind the other in a showroom.

1) If the positions of cars C3 and C4 are interchanged, then the positions of cars before C3 and cars after C4 will change.
   A) True   B) False

2) If car C3 is removed from the line-up, then the fifth car in the line up will be
   A) C5   B) C6

10

When Joe was in the movie theater, he saw advertisements in the following order: Soap, Popcorn, Juice, Shoes

1) What is the first food related advertisement that he saw?

2) What is the last food related advertisement that he saw?

3) Which non-food related advertisement was shown after the advertisement for popcorn?

4) Which non-food related advertisement was shown before the advertisement for shoes?

5) Which advertisement did Joe see between the Soap and Juice advertisements?

Name _____ Date _____

SEQUENCING - Order

11
ascending

Sam picked up carts numbered 9, 3, 6, and 2 and brought them into the store.

The sequence of the carts that were brought into the store in ascending order is

    A) 2,3,6,9
    B) 9,3,6,2

12
top to bottom

There were several chairs in the lawn. Stephanie stacked them on top of each other with the lowest numbered chair at the bottom.

Which of the following sequence of chairs listed from the top is correct?

    A) 1, 2, 3, 4
    B) 1, 4, 3, 2
    C) 5, 3, 2, 1

Analytical Reasoning   Answers-113

SEQUENCING - Symbolic

13

Symbol [→1] means take 1 step forward and 2 steps backward.
Symbol [→2] means take 2 steps forward and 1 step backward.
Symbol [←1] means take 1 step backward and 2 steps forward.
Symbol [←2] means take 2 steps backward and 1 step forward.

1) If the instructions in the symbols shown above are followed, then what does the following sequence of symbols mean taken together?

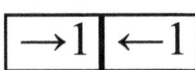

A) take 1 step backward
B) take 1 step forward
C) take 2 steps forward
D) no change in position

2) [→2] and [←1] effectively mean the same.

   A) True   B) False

3) Draw a sequence of symbols that will take someone five steps forward.

SEQUENCING - Project Plan

14

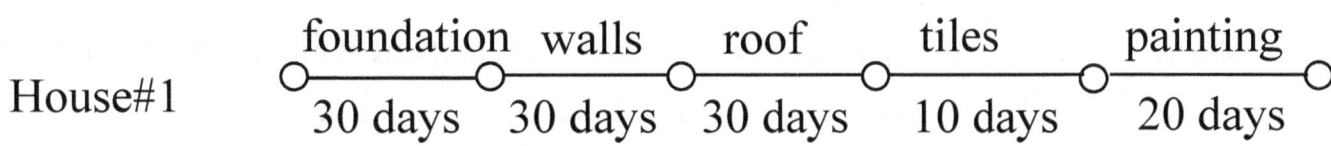

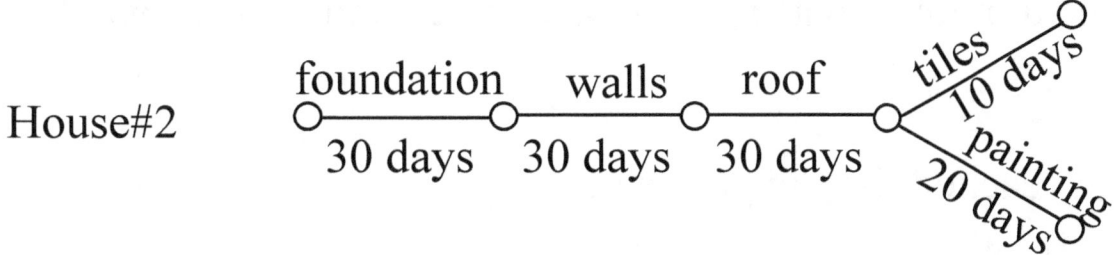

Construction plans of two houses are shown above. The painting task and the tiles task in the plan for house# 2 can be done simultaneously. Read the plans and answer the questions below.

1) If construction for both houses begins on the same day, then the construction for both houses will be completed on the same day.
    A) True   B) False

2) If construction for both houses begins on the same day, then the tile laying task for both houses will be completed on the same day.
    A) True   B) False

3) Painting for house#2 will start before painting for house# 1.
    A) True   B) False

Name _____   Date_____

SEQUENCING - Timing

15

Three boats with two people each in them participated in a speed boating competition. The boats left the starting line consecutively at 8 AM (boat#1), 9 AM (boat#2) and 10 AM (boat#3). They reached the finishing line at noon (boat#1), 11 AM(boat#2) and 1 PM(boat#3) respectively.

1) The first boat reached the finish line first.
    A) True B) False

2) Which of the following must be true?
  A) The boat that arrived at the finish line first must be the winner.
  B) The boat that takes the least time to reach the finish line is the winner.

3) The order of the boats in terms of how long they took to reach the finish line, from least to most is
    A) boat#3, boat#2, boat#1
    B) boat#1, boat#2, boat#3
    C) boat#2, boat#3, boat#1

Analytical Reasoning   Answers-115
© Gift Of Logic, Inc * Copying prohibited

Name _____   Date _____

SCHEDULING

1 A cardiologist sees patients every other day starting from the first of the month. Answer the questions below for the month shown.

1) The cardiologist did not see patients on Saturdays.
    A) True   B) False

2) The cardiologist did not see patients on Sundays.
    A) True   B) False

---

2 After an accident involving a bus, patients were taken to a hospital in a line. The doctors there will attend to them on a first-come, first-served basis. Josh was at the first to arrive at the hospital and Jane was the last to arrive. Jamie also was taken to the hospital.

1) Jamie will be scheduled to be seen before Josh.
    A) True   B) False

2) Jamie will be scheduled to be seen before Jane.
    A) True   B) False

Name _____   Date _____

SCHEDULING - constraints

3

A hospital has to schedule three doctors, A, B, and C to take care of patients from Monday-Friday.

Doctor A can work on Monday and Wednesday, but not on other days.
Doctor B can work on Monday and Tuesday, but not on other days.
Doctor C can work on Thursday and Friday only.
A doctor can work at most two days only.

Which of the following schedules of the three doctors is correct? Mark your answers with a check mark in the column titled "Correct".

Monday	Tuesday	Wednesday	Thursday	Friday	Correct
A	B	A	C	C	
B	B	A	C	C	
A	B	C	A	B	

4

Doctor Even works on even numbered days.
Doctor Odd works on odd numbered days.

1) Since doctors Even and Odd take alternate turns to work, they work the same number of days each month.
   A) True   B) False

2) Doctor Even works on the last day of April, June, September and November.
   A) True   B) False

Name _____   Date_____

STRATEGY PROBLEMS - Exchanging

In the following strategy problems, you will learn to select the correct strategy to find a solution to the given problem.

1

There are seven people in team A and three people in team B.

What is the correct strategy to make both teams have the same number of people?

    A) move 2 people form team A to team B.
    B) move 2 people from team B to team A.

---

2   There are 4 boys and 2 girls in team Orange and 4 girls and 2 boys in team Apple.

What is the correct strategy to have the same number of boys and girls in each team?

A) move 1 boy from team Orange to team Apple and 1 girl from team Orange to team Apple.

B) move 1 boy from team Orange to team Apple and 1 girl from team Apple to team Orange.

C) move 1 girl from team Orange to team Apple and 1 boy from team Apple to team Orange.

Analytical Reasoning   Answers-118
© Gift Of Logic, Inc * Copying prohibited

Name _____   Date _____

STRATEGY PROBLEMS - Exchanging

3

Gardener Graham has made a mistake. He planted plants A, B, and C in garden A and plants D, E, and F in garden B, but plant A will not grow in garden A and plant E will not grow in garden B. Moreover, plants B and E cannot grow together in garden A, and plants A and D will not grow together in garden B.

Please help devise a strategy for Gardener Graham so that all the plants can grow.

A) → move plant A from garden A to garden B.
    → move plant E from garden B to garden A.

B) → move plant A from garden A to garden B.
    → move plant E from garden B to garden A.
    → move plant C from garden A to garden B.
    → move plant F from garden B to garden A.

C) → move plant A from garden A to garden B.
    → move plant E from garden B to garden A.
    → move plant B from garden A to garden B.
    → move plant D from garden B to garden A.

Analytical Reasoning   Answers-119
© Gift Of Logic, Inc * Copying prohibited

Name ——————————————   Date ————————————

STRATEGY PROBLEMS - Timing

4

Jenny has three homeworks to do starting at 8 AM. Homework-1 will take 2 hours, homework-2 will take 4 hours, and homework-3 will take 2 hours. Each homework has to be done in one sitting without a break. They can be done in any order. She must take a 1 hour break starting at noon.

What is the best strategy for Jenny to complete her homeworks?
  A) do homeworks 1 and 2 before noon and then, do homework-3.
  B) do homeworks 2 and 3 before noon and then, do homework-1.
  C) do homeworks 1 and 3 before noon and then, do homework-2.

5

Lee is planing to visit his sister and uncle who live in different parts of the town. It will take 30 minutes each way to go and come back from his sister's house, and 20 minutes each way to go and come back from his uncle's house. From his sister's house, it will take 20 minutes to go to his uncle's house, and from his uncle's house it will take 40 minutes to go to his sister's house.

What is the best strategy for Lee to visit both his sister and uncle at the shortest time possible?
    A) go to his sister's house first, and then to his uncle's.
    B) go to his uncle's house first, and then to his sister's.

Analytical Reasoning   Answers-120

Name _____   Date _____

STRATEGY PROBLEMS - Mixing

6

There are four liquids A, B, C, and D. B and C when mixed, will cause a loud noise. A and D when mixed, will cause fire.

What is the strategy to make a compound solution consisting of three of these liquids that will not cause a loud noise?

    A) mix A, D, and B
    B) mix B, D, and C

Is there any other 3-solution compound that will not cause a loud noise?

7

Vaccine-X can kill germs A and B, but not C.
Vaccine-Y can kill germs B and C, but not A.
Vaccine-Z can kill germs C and A, but not B.

During the winter season, all the three germs struck a city. Which of the following is the best strategy for a Doctor to kill all the germs with the least number of vaccines?

    A) give one vaccine only
    B) give any two vaccines
    C) give all the vaccines

Name _____   Date _____

STRATEGY PROBLEMS - Timing

8

At 4 PM, the pilots of a Jumbo jet were told that there will be bad weather for landing for the next 5 hours. At that time, the airplane had 4 hours worth of fuel left.

The best strategy for the pilot is to

  A) land the airplane after the bad weather improves.
  B) land the airplane in bad weather.

9

Jack can take either a 3-hour or a 4-hour flight at 1 PM to Denver. Jack's connecting flight at Denver is at 6 PM. In order to have the least stopover time at Denver before catching the connecting flight, the best strategy for Jack would be to

  A) take the 3-hour flight.
  B) take the 4-hour flight.

Analytical Reasoning   Answers-122

Name _____   Date _____

STRATEGY PROBLEMS - maximum, minimum

10 Ferry#1 and Ferry#2 carry passengers from island A to islands B, C and D. The ferry departure timings at the islands are shown below. The ferries arrive at the islands five minutes before the indicated departure time.

Island	Ferry#1	Ferry#2
A	8 AM	8:30 AM
B	9 AM	9:30 AM
C	11 AM	10:30 AM
D	11:30 AM	11:30 AM

Martin, currently at island A, has to attend a meeting from 9 AM to 9:15 AM at island B, and a meeting at 11:45 AM at island D. The best strategy for Martin to attend the meetings is
    A) take Ferry#1.
    B) take Ferry#1 and Ferry#2.

---

11 The Mayor of Clean City is mad because too much garbage has piled up on the city streets. He ordered the city manager to remove the garbage as soon as possible.

Assuming that all the trucks are of the same size, the best strategy for the city manager to remove the garbage quickly is
    A) to use maximum number of trucks possible to haul the garbage.
    B) to use the minimum number of trucks possible to haul the garbage.

Assuming that all the trucks are of the same size, the best strategy for the city manager to remove the garbage with the least expense is
    A) to use maximum number of trucks to haul the garbage.
    B) to use the minimum number of trucks to haul the garbage.

Analytical Reasoning   Answers-122

STRATEGY PROBLEMS - Quickest, Shortest

12

Ali was going to work in his car when he realized that he had enough fuel only to go for another 20 miles. His office is 10 minutes away at 22 miles. If he takes a detour, he can reach his office in 20 minutes after traveling 18 miles.

The best strategy for Ali to reach his office without running out of fuel is to
    A) not take the detour.
    B) take the detour.

13

Goldstein has to deliver furniture to his customer. He can choose to drive on one of two roads - a 40 mile road or 60 mile road. The 40 mile road takes 4 hours to drive, but the 60 mile road takes 2 hours only because of higher speed limits.

1) To deliver the furniture quickly, Goldstein should take
    A) the 40 mile road.
    B) the 60 mile road.

2) Goldstein gets rewarded for driving the shortest distance. To get the maximum reward, Goldstein must take
    A) the 40 mile road.
    B) the 60 mile road.

Analytical Reasoning   Answers-124

| 1 | POSITIONING PROBLEMS - Vacancy |

Three butterflies fly into a tree branch. There are four spots on the branch where they can possibly sit as shown below. Only one butterfly can sit in one spot.

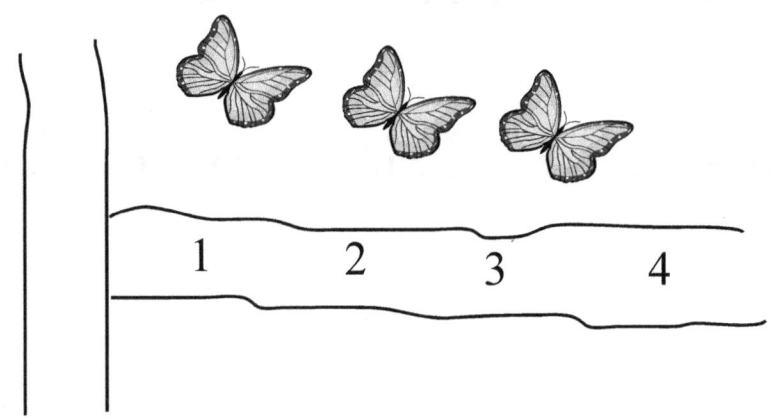

1) After all the butterflies sit on the branch, how many spots will be without a butterfly?
    A) 1   B) 2   C) 3   D) 4

2) If two butterflies sit in spots 1 and 3 respectively, then which of the following spots will be vacant?
    A) 2 and 4   B) 2 or 4, but not both

3) Regardless of where the three butterflies sit, there will always be one vacant spot.
    A) True   B) False

Analytical Reasoning   Answers-125
© Gift Of Logic, Inc * Copying prohibited

2	POSITIONING PROBLEMS - Vacancy

Three butterflies fly in to a tree branch. There are four spots on the branch where they can possibly sit as shown below. Only one butterfly can sit in one spot.

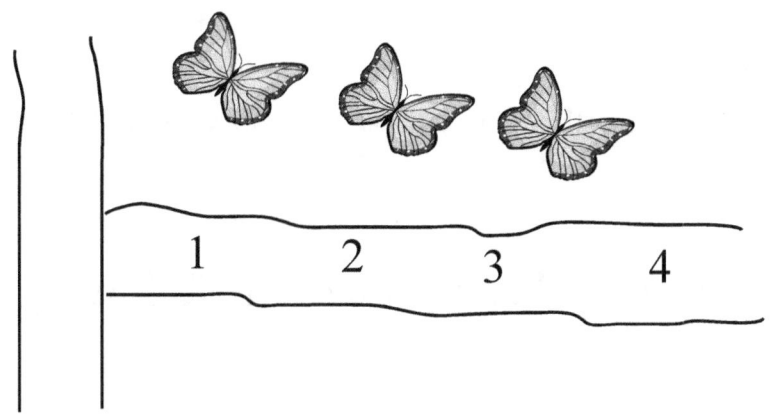

1) If all the three butterflies must sit next to each other, then which of the following spots can be vacant?

A) spots 1 or 4   B) spots 2 or 3

2) If no butterfly can sit in spot 3, then which one of the following must be true?
   A) There will be one butterfly to the left of spot 3.
   B) There will be two butterflies to the left of spot 3.

3   POSITIONING PROBLEMS - Vacancy

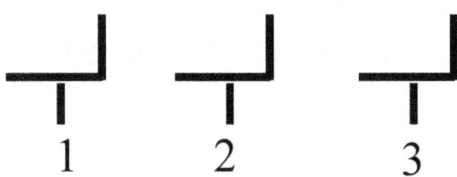

Two boys Ying and Zac must be seated in the three chairs shown above, one in each chair.

1) If Ying must sit to the left of Zac, then which of the following seating arrangements is possible?

1)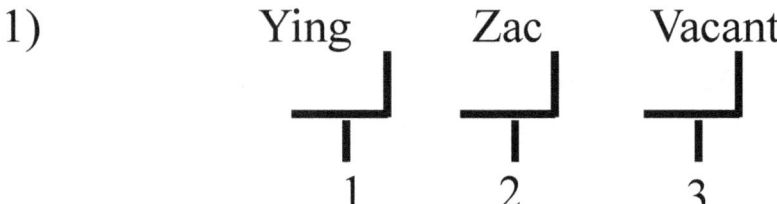

2)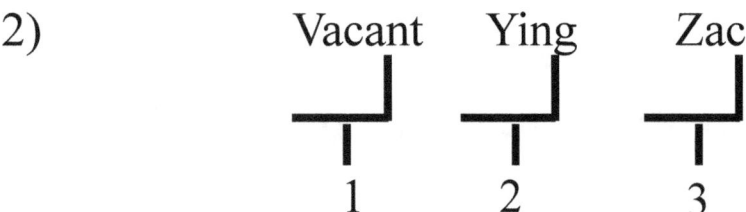

3) Ying   Vacant   Zac

| 4 | POSITIONING PROBLEMS - No vacancy |

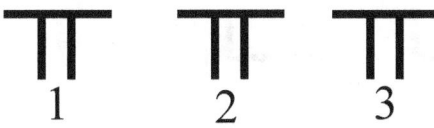

Four girls A, B, C, and D are to be seated in the three stools shown above. Only one girl in each stool. If A gets a stool, then D must also get a stool.

1) How may students will be left without a stool?
   A) 1   B) 2   C) 3

2) If A and B are seated, then who will not be able to get a stool?
   A) D   B) C

3) Which one of the following statements must be true?
   A) A, B, and C can be seated in the stools.
   B) D, B, and C can be seated in the stools.

Analytical Reasoning   Answers-128

5 POSITIONING PROBLEMS - No vacancy

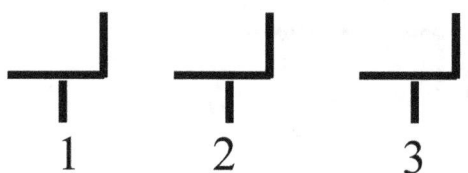

Five boys A, B, C, D, and E are to be seated in three chairs shown above. Only one boy in each chair. If A gets a seat, then C must also get a seat. If B gets a seat, then D must also get a seat.

1) How may students will be left without a chair?
   A) 1   B) 2   C) 3

2) If A and D are seated, then who will not be able to get a seat?
   A) B and E   B) C and E

3) Which one of the following statements must be true?
   A) C, B, and E can be seated in the chairs.
   B) C, D, and E can be seated in the chairs.

Name _____ Date_____

1   PICKING PROBLEMS

Three balls must be selected from a set of five available balls shown above.

If Red is picked, then Green must be picked.
If Green is picked, Blue must be picked.

1) Which of the following picks is correct?
   A) Red, Green, White
   B) Green, Blue, White

2) If Red is picked, then White and Pink will not be picked.
   A) True   B) False

3) White and Pink cannot be picked together.
   A) True   B) False

4) How many balls will always not be picked?
   A) 2   B) 3

Analytical Reasoning   Answers-130
© Gift Of Logic, Inc * Copying prohibited

2   PICKING PROBLEMS

Box A
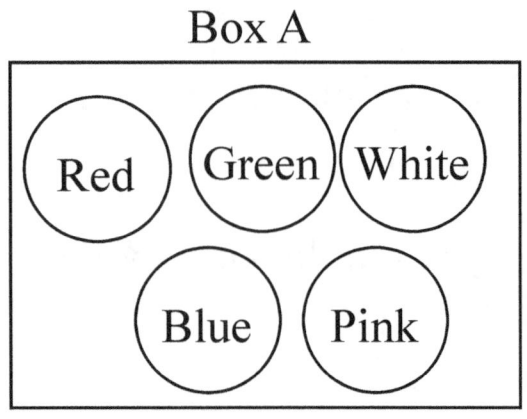

Box B
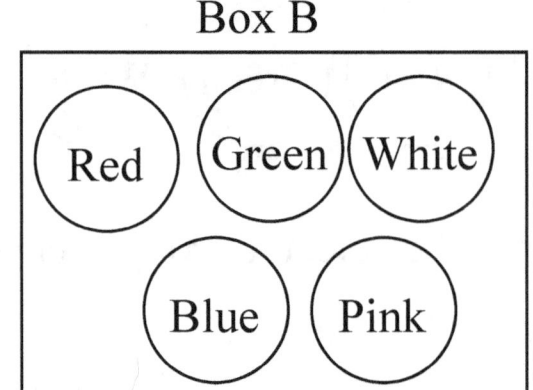

Two boxes contain balls of different colors as shown. A total of five balls must be selected from these two boxes. If one red ball is selected, two green balls must also be selected. If one green ball is selected, then two blue balls must be selected.

1) Which of the following statements must be true?
   A) A Red ball cannot be selected.
   B) A green ball can be selected without a blue ball.

2) Which one of the following selections is correct?
   A) Blue, Blue, White, Pink, Green.
   B) Green, Blue, White, Pink, Green.

Name ____________________ Date____________________

NUMERIC SUDOKU

Solve the Sudokus shown below. A solved Sudoku has numbers 1,2,3, and 4 appearing in each row, each column and the four bolded squares only once.You develop valuable positioning skills while solving these Sudokus.

1

1		2	
	2		
4		3	
		1	

2

4			3
		4	
3		1	
			2

| Name _____ | Date _____ |

NUMERIC SUDOKU

Solve the Sudokus shown below. A solved Sudoku has numbers 1,2,3, and 4 appearing in each row, each column and the four bolded squares only once. You develop valuable positioning skills while solving these Sudokus.

3

		2	
1			3
	4	1	

4

4			
		4	2
	4	3	
3	1		

Name ____________________ Date____________________

NUMERIC SUDOKU

Solve the Sudokus shown below. A solved Sudoku has numbers 1,2,3, and 4 appearing in each row, each column and the four bolded squares only once. You develop valuable positioning skills while solving these Sudokus.

5

2			3
3			
	2		4
		1	

6

	3		1
2		4	
	2		
1			2

Name _____   Date_____

NUMERIC SUDOKU

Solve the Sudokus shown below. A solved Sudoku has numbers 1,2,3, and 4 appearing in each row, each column and the four bolded squares only once. You develop valuable positioning skills while solving these Sudokus.

7

	3	2	
	1		
4		3	

8

3	4		
		3	
1	3	4	2

NUMERIC SUDOKU

Solve the Sudokus shown below. A solved Sudoku has numbers 1,2,3, and 4 appearing in each row, each column and the four bolded squares only once. You develop valuable positioning skills while solving these Sudokus.

9

	1	2	
			4
	2		1

10

1			4
4		1	
		4	
3		2	

NUMERIC SUDOKU

Solve the Sudokus shown below. A solved Sudoku has numbers 1,2,3, and 4 appearing in each row, each column and the four bolded squares only once. You develop valuable positioning skills while solving these Sudokus.

11

2			3
		2	1
	2		4
3			

12

1			3
		2	
3		1	
	1		4

ALPHABETIC SUDOKU

Solve the Sudokus shown below. A solved Sudoku has alphabets A,B,C, and D appearing in each row, each column and the four bolded squares only once. You develop valuable positioning skills while solving these Sudokus.

1

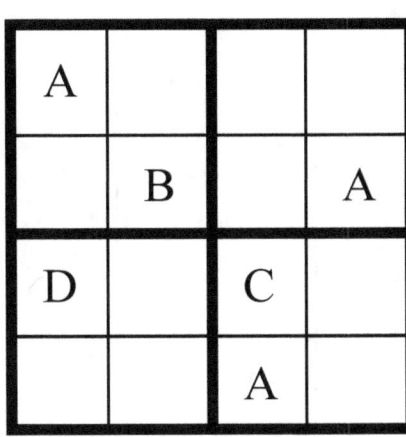

2

D			
B		D	
C		A	
			B

Analytical Reasoning   Answers-138

Name _____   Date_____

ALPHABETIC SUDOKU

Solve the Sudokus shown below. A solved Sudoku has alphabets A, B, C, and D appearing in each row, each column and the four bolded squares only once. You develop valuable positioning skills while solving these Sudokus.

3

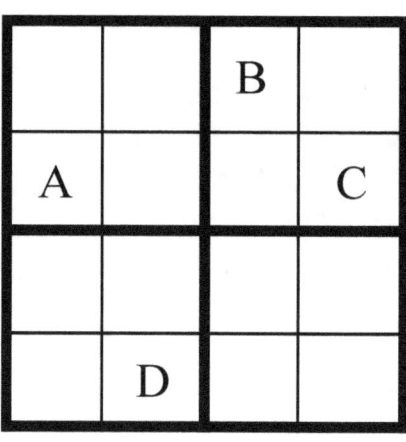

4

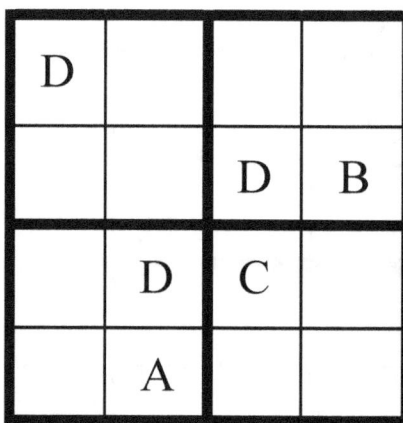

Analytical Reasoning   Answers-139
© Gift Of Logic, Inc * Copying prohibited

ALPHABETIC SUDOKU

Solve the Sudokus shown below. A solved Sudoku has alphabets A, B, C, and D appearing in each row, each column and the four bolded squares only once. You develop valuable positioning skills while solving these Sudokus.

5

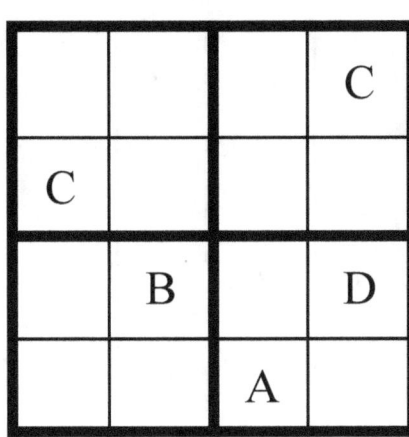

6

ALPHABETIC SUDOKU

Solve the Sudokus shown below. A solved Sudoku has alphabets A, B, C, and D appearing in each row, each column and the four bolded squares only once. You develop valuable positioning skills while solving these Sudokus.

7

	C	B	
	A		
D		C	

8

	D		
		C	
		A	
A	C		

ALPHABETIC SUDOKU

Solve the Sudokus shown below. A solved Sudoku has alphabets A,B,C, and D appearing in each row, each column and the four bolded squares only once. You develop valuable positioning skills while solving these Sudokus.

9

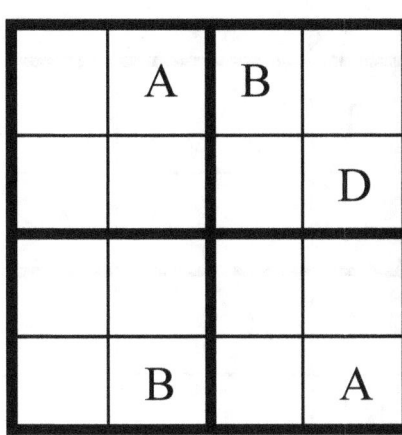

10

ALPHABETIC SUDOKU

Solve the Sudokus shown below. A solved Sudoku has alphabets A,B,C, and D appearing in each row, each column and the four bolded squares only once. You develop valuable positioning skills while solving these Sudokus.

11

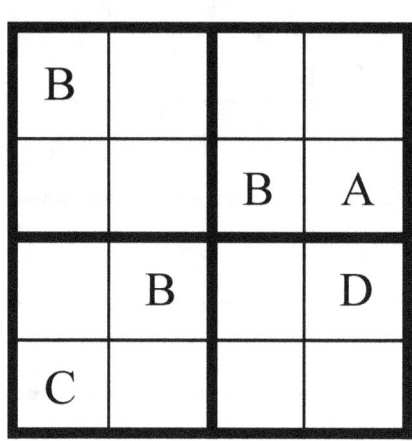

12

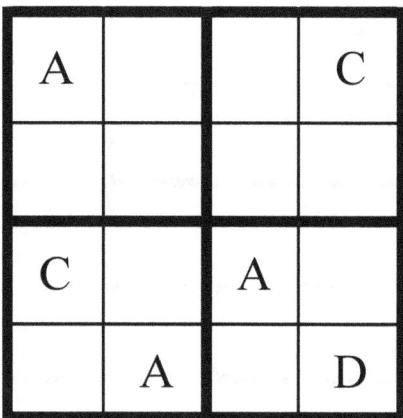

PICTORIAL REASONING

Name _____   Date _____

PICTURE SEQUENCE

Find the logic in the sequence of figures shown, and pick the correct figure from the choices given that will continue the sequence.

1
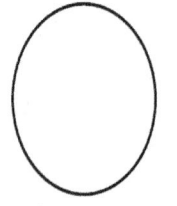 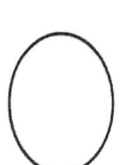 ?   A    B

2
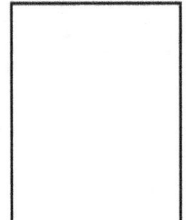 ?   A   B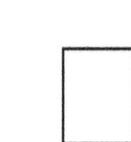

3
 ?   A 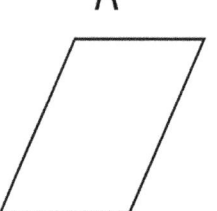   B

4
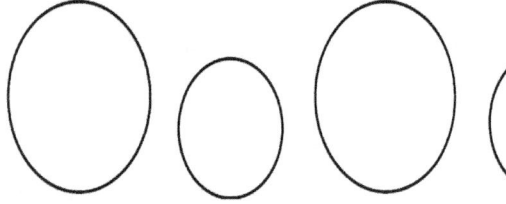 ?   A 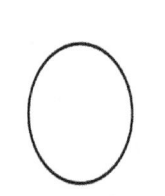   B

Pictorial Reasoning   Answers-144

© Gift Of Logic, Inc * Copying prohibited

Name _____   Date _____

PICTURE SEQUENCE

Find the logic in the sequence of figures shown, and pick the correct figure from the choices given that will continue the sequence.

5   A   B

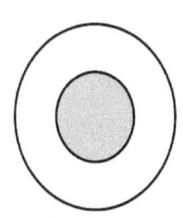 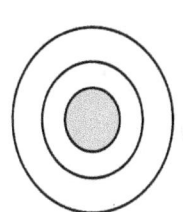 ?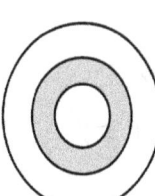

6   A   B

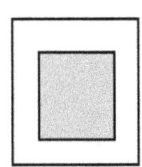 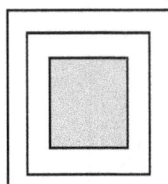 ?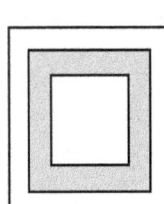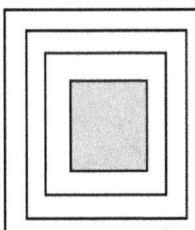

7   A   B

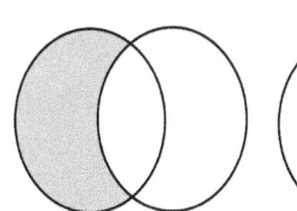 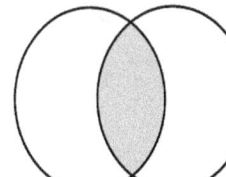 ?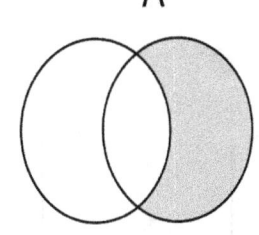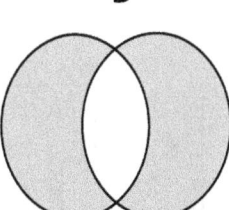

8   A   B

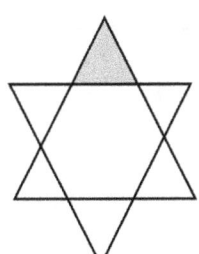 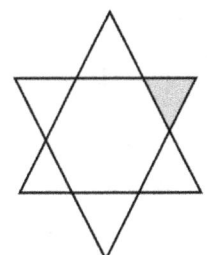 ?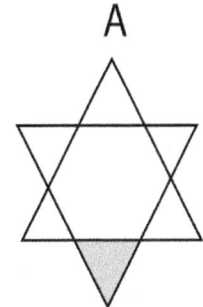

Pictorial Reasoning   Answers-144
© Gift Of Logic, Inc * Copying prohibited

Name _____ Date_____

PICTURE SEQUENCE

Find the logic in the sequence of figures shown, and pick the correct figure from the choices given that will continue the sequence.

9   A   B

 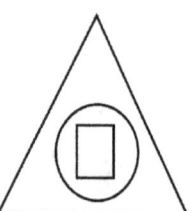   ?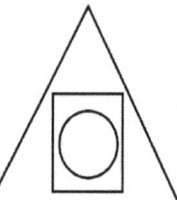

10   A   B

 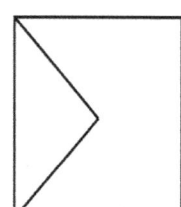   ?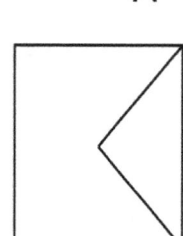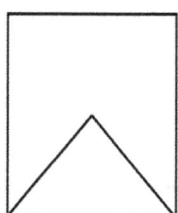

11   A   B

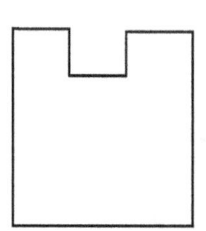 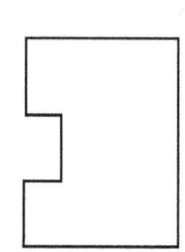   ?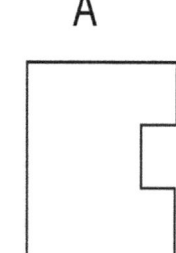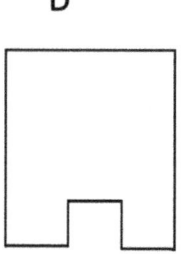

12   A   B

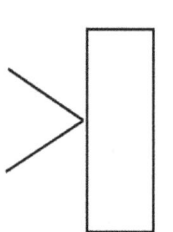 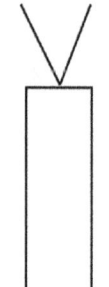   ?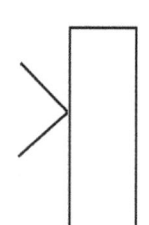

Pictorial Reasoning   Answers-144
© Gift Of Logic, Inc * Copying prohibited

Name _____ Date _____

PICTURE SEQUENCE

Find the logic in the sequence of figures shown, and pick the correct figure from the choices given that will continue the sequence.

13   A   B

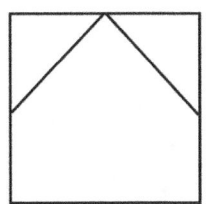 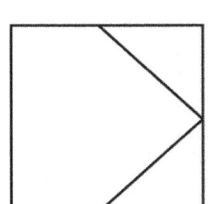 ?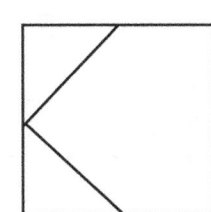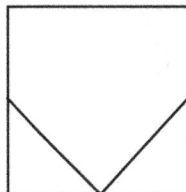

14   A   B

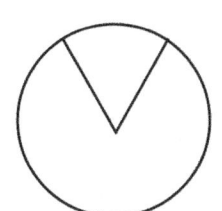 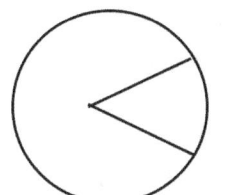 ?

15   A   B

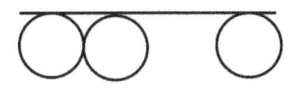 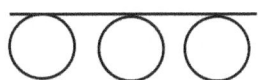 ?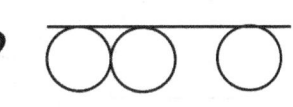

16   A   B

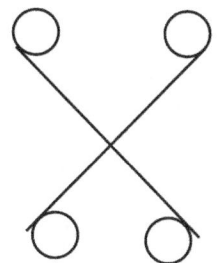 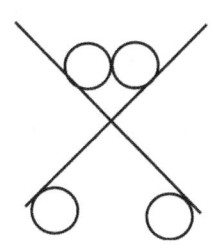 ?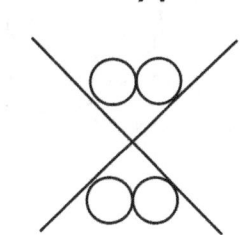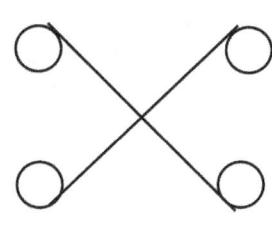

Pictorial Reasoning   Answers-144   75
© Gift Of Logic, Inc * Copying prohibited

Name _____   Date _____

PICTURE DIFFERENCE

Mark the difference between the two pictures in each set.

1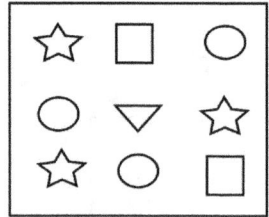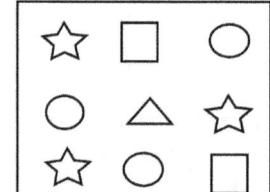

2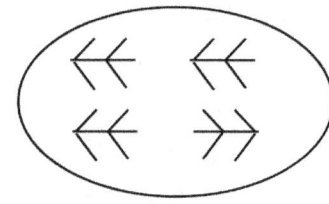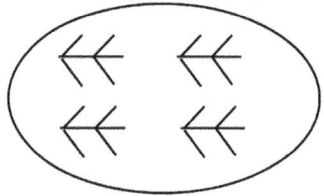

3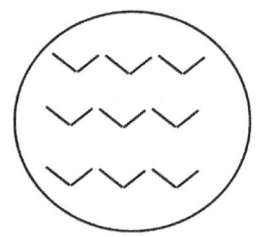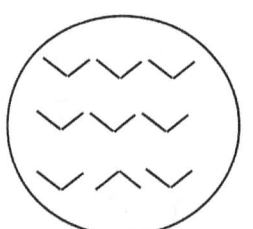

4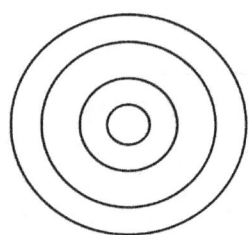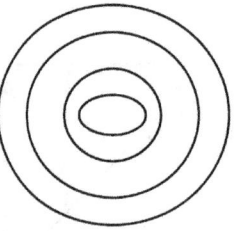

Name _____ Date _____

PICTURE DIFFERENCE

Mark the difference between the two pictures in each set.

5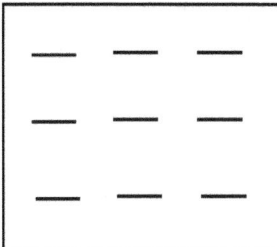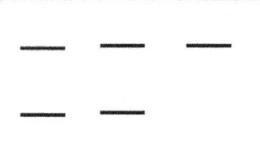

6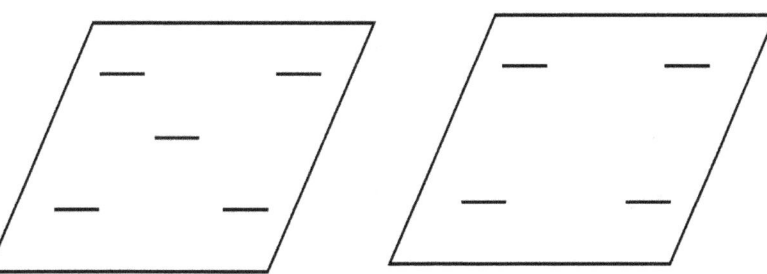

7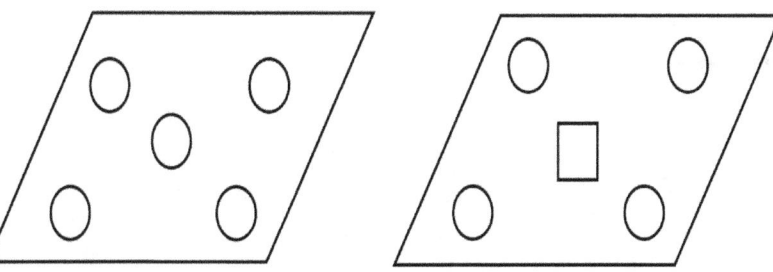

8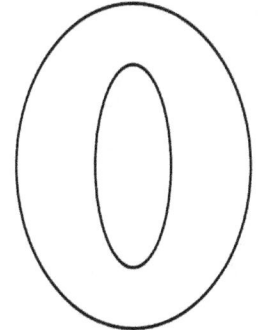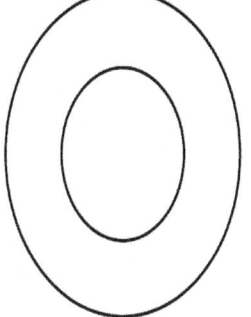

Pictorial Reasoning   Answers-145
© Gift Of Logic, Inc * Copying prohibited

Name _____ Date _____

PICTURE DIFFERENCE

Mark the difference between the two pictures in each set.

9

10

11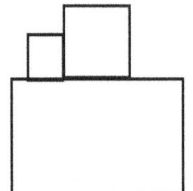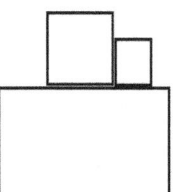

12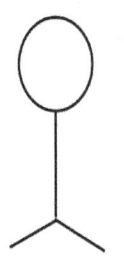

Pictorial Reasoning   Answers-145
© Gift Of Logic, Inc * Copying prohibited

Name _____ Date _____

PICTURE DIFFERENCE

Mark the difference between the two pictures in each set.

13

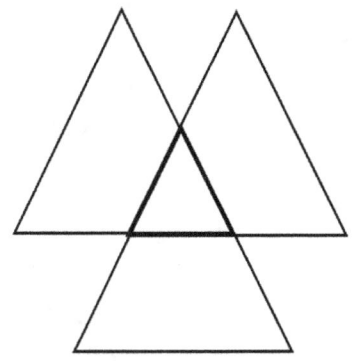 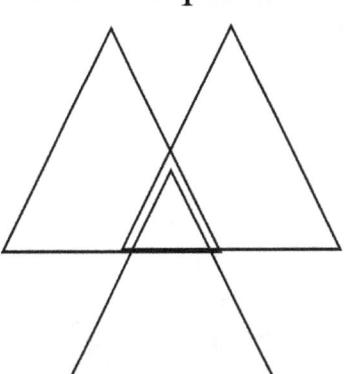

14

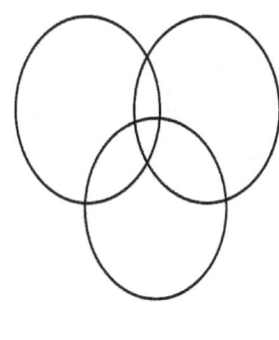 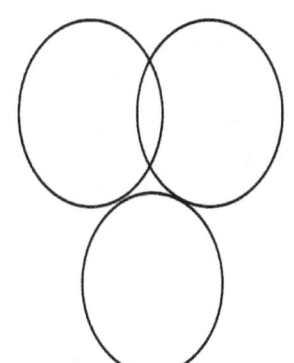

15

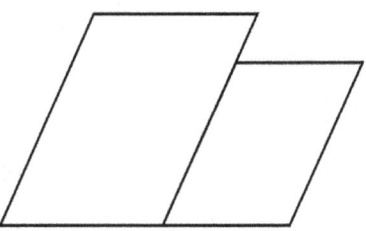 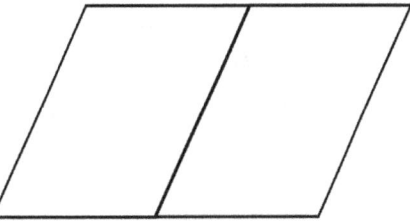

16

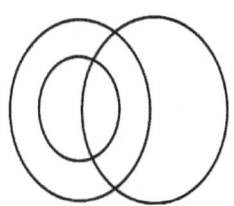 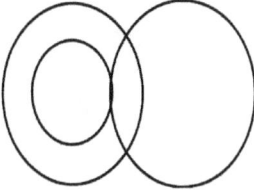

Pictorial Reasoning   Answers-145
© Gift Of Logic, Inc * Copying prohibited

Name _____   Date _____

PICTURE ANALOGY

Circle the correct choice (A or B) that will complete the picture analogy.

1
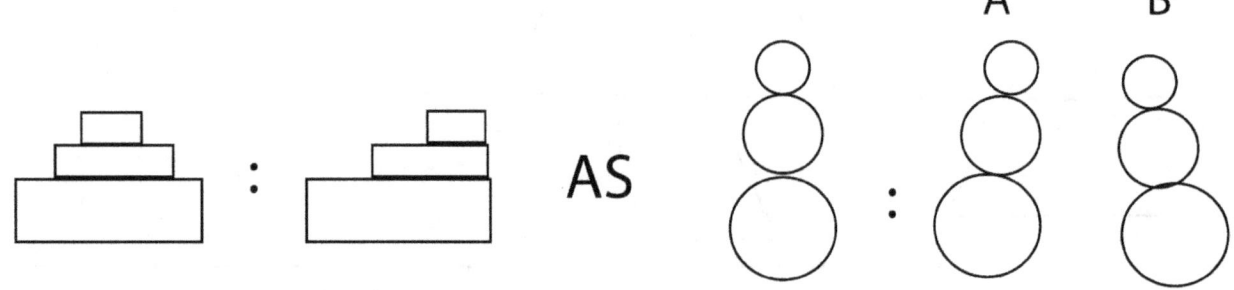

2
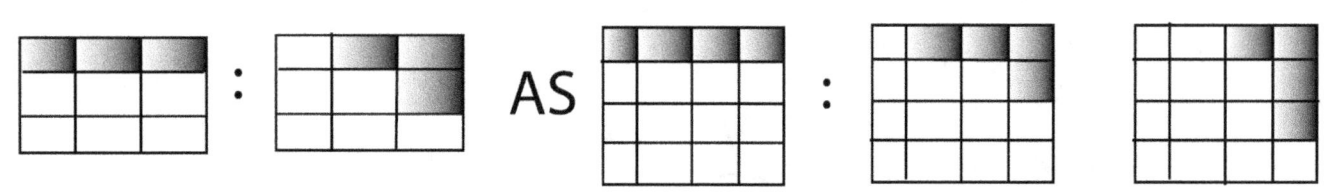

3

4
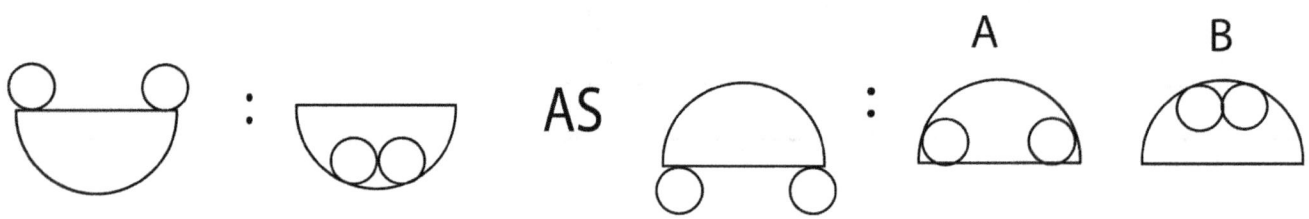

Pictorial Reasoning   Answers-146
© Gift Of Logic, Inc * Copying prohibited

Name _____   Date _____

PICTURE ANALOGY

Circle the correct choice (A or B) that will complete the picture analogy.

5

6

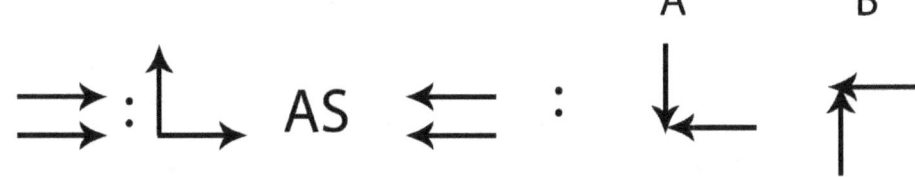

7

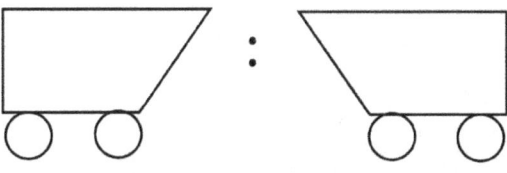 AS  :

8

 :  AS

Pictorial Reasoning   Answers-146
© Gift Of Logic, Inc * Copying prohibited

Name _____   Date _____

PICTURE ANALOGY

Circle the correct choice (A or B) that will complete the picture analogy.

9   A   B

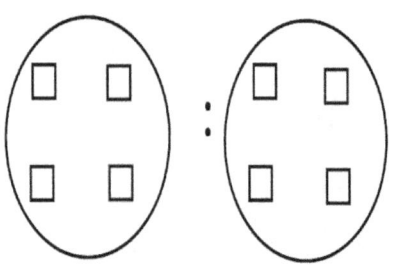 AS 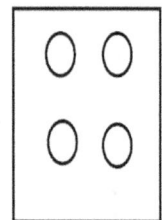 :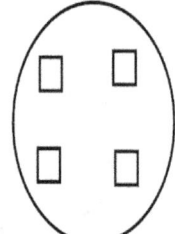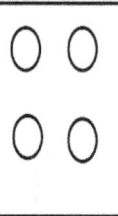

10   A   B

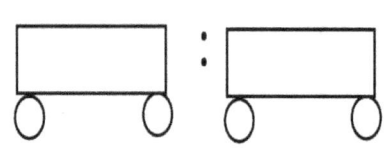 AS 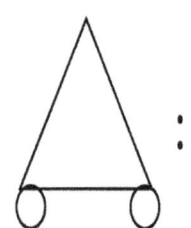 :

11   A   B

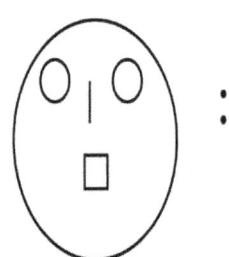  AS 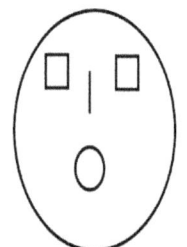 :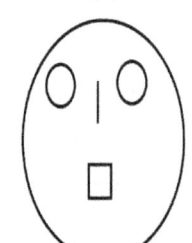

12   A   B

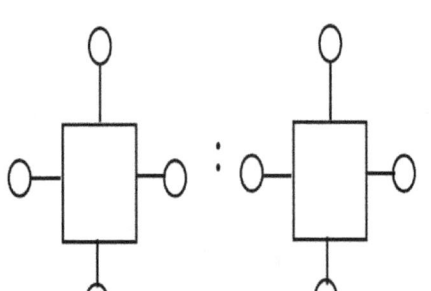 AS 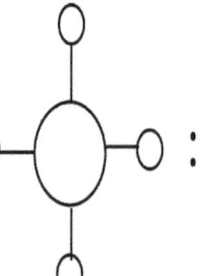 :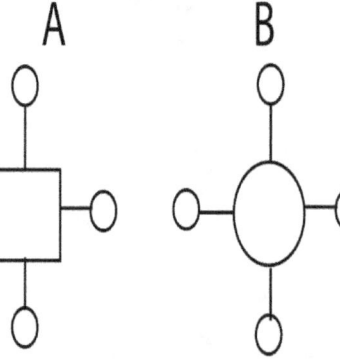

Pictorial Reasoning   Answers-146
© Gift Of Logic, Inc * Copying prohibited

Name _____ Date _____

ODD PICTURE

Circle the odd picture in each question below based on its pattern.

1   A   B   C

 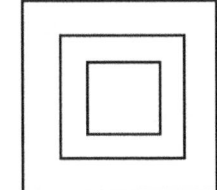 

2   A   B   C

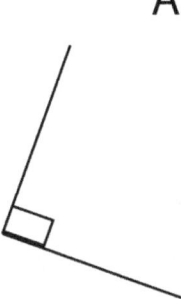   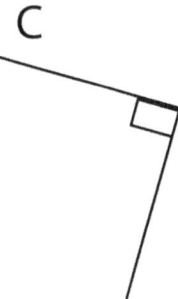

3   A   B   C

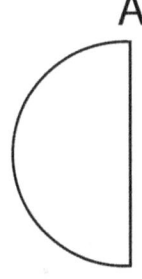 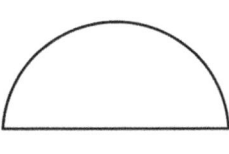 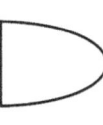

4   A   B   C

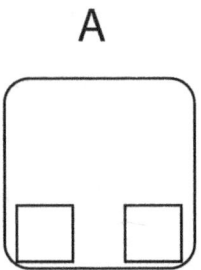 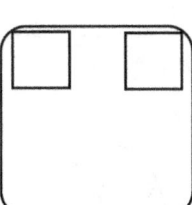 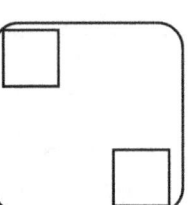

Pictorial Reasoning   Answers-147
© Gift Of Logic, Inc * Copying prohibited

Name _____ Date _____

ODD PICTURE

Circle the odd picture in each question below based on its pattern.

5   A   B   C

6   A   B   C

  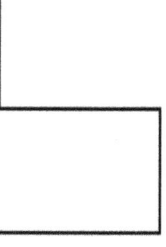

7   A   B   C

8   A   B   C

 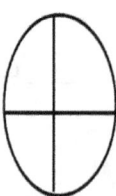 

Pictorial Reasoning   Answers-147
© Gift Of Logic, Inc * Copying prohibited

Name —————————————— Date ——————————————

ODD PICTURE

Circle the odd picture in each question below based on its pattern.

9   A   B   C

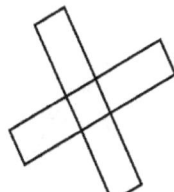 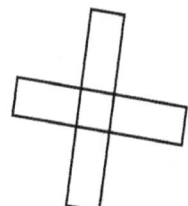 

10   A   B   C

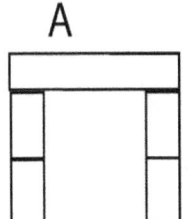 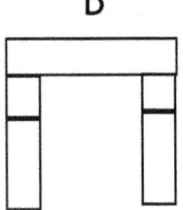 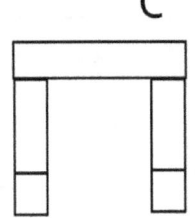

11   A   B   C

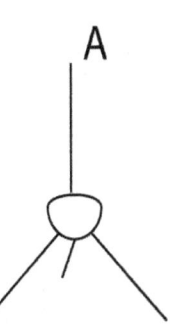 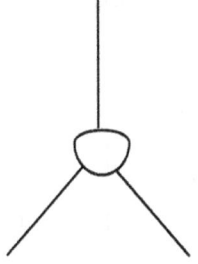 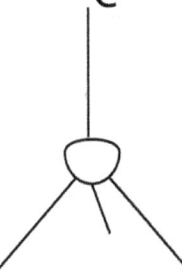

12   A   B   C

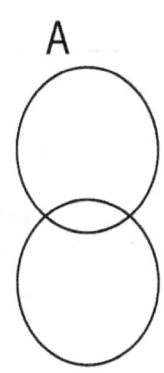 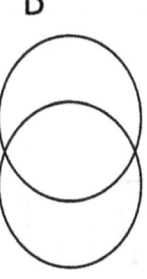 

Pictorial Reasoning   Answers-147
© Gift Of Logic, Inc * Copying prohibited

Name _____   Date _____

ODD PICTURE

Circle the odd picture set in each question below based on its pattern.

13   A   B   C

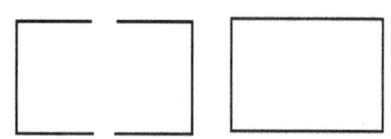 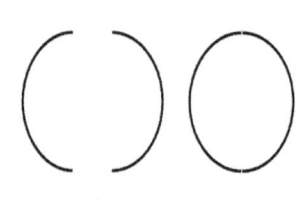 

14   A   B   C

 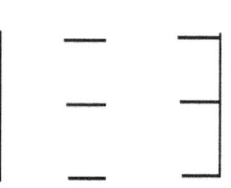 

15   A   B   C

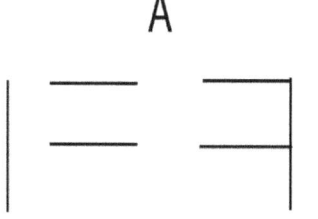  

16   A   B   C

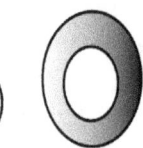 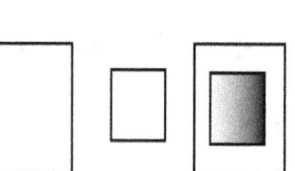 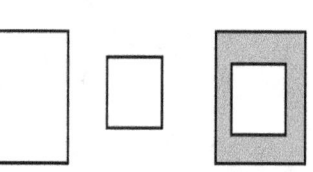

Pictorial Reasoning   Answers-147
© Gift Of Logic, Inc * Copying prohibited

Name _____   Date _____

PATTERN MATCHING

Find the logical pattern in the pictures on the left, and identify the picture on the right that will fit in the space marked with ? to complete the pattern.

1
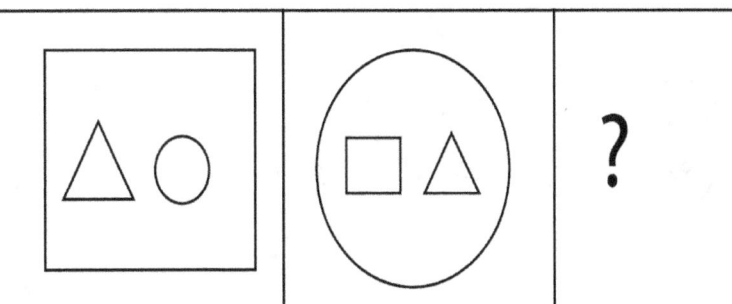 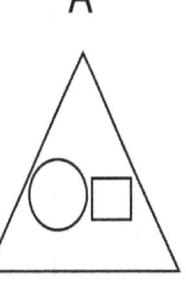 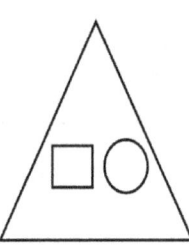

2
 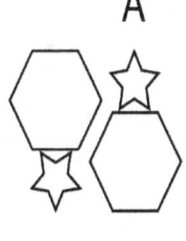 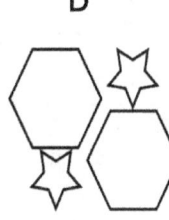

3
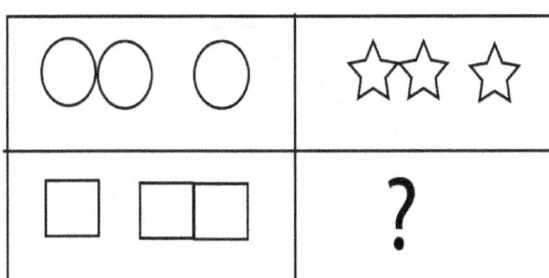 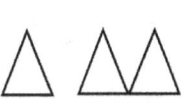 

4

Pictorial Reasoning   Answers-148
© Gift Of Logic, Inc * Copying prohibited

Name _____   Date _____

PATTERN MATCHING

Find the logical pattern in the pictures on the left, and identify the picture on the right that will fit in the space marked with ? to complete the pattern.

5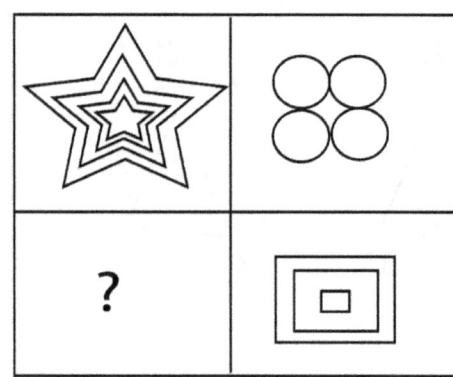

6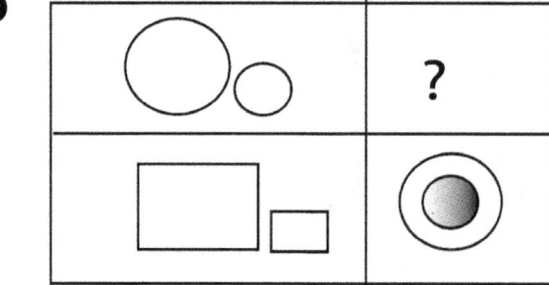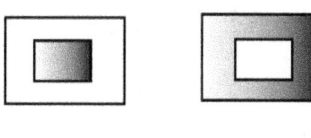

7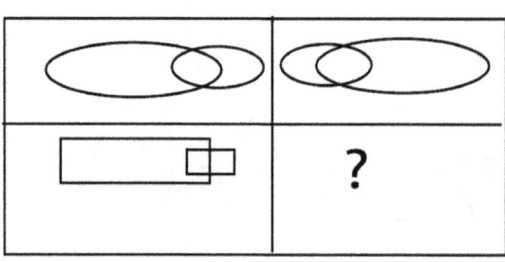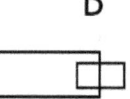

8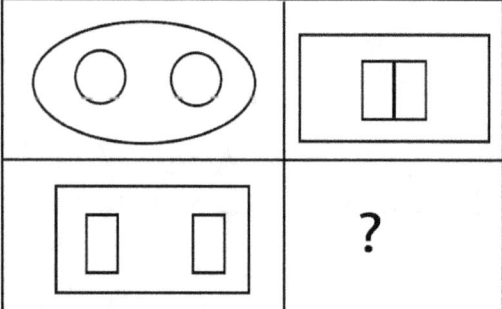

Pictorial Reasoning   Answers-148
© Gift Of Logic, Inc * Copying prohibited

Name _____   Date _____

PATTERN MATCHING

Find the logical pattern in the pictures on the left, and identify the picture on the right that will fit in the space marked with ? to complete the pattern.

9 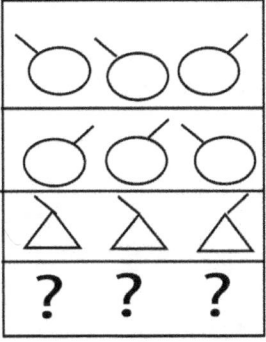   A   B
△△△   △△△

10 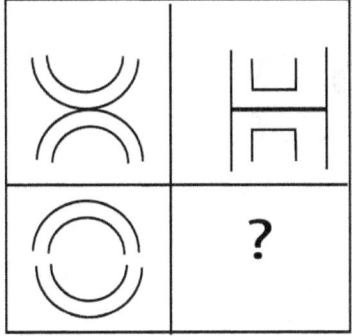   A   B

11 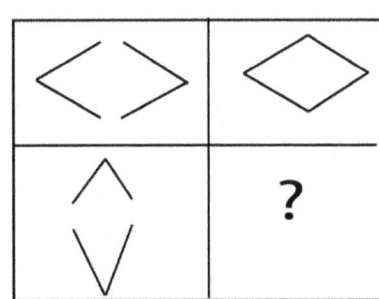   A   B
◇   ◇

12 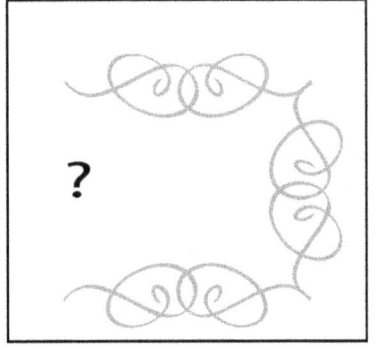   A   B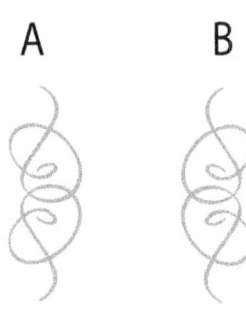

Pictorial Reasoning   Answers-148
© Gift Of Logic, Inc * Copying prohibited

ANSWERS

FINDING THE TRUTH

Q#	Answer	Reasoning
1	Varies	For this statement to be true, you should wake up before 7 AM on all Tuesdays, not just some Tuesdays.
2	Varies	Could be true or false. Think how this statement could be true for someone, and false for others.
3	Varies	Could be true or false. For this statement to be true, you should play chess on all days except on Wednesdays.
4	B-False	Planet Mars also orbits around the Sun.
5	B-False	Exclusive means restricted. This means that John can not bike in a lane reserved for cars only.
6	B-False	Since the game is exclusively shown in channel 19, it can not be seen in any other channel.
7	A-True	Amphibians are animals that live in water and land. Turtles can live in water and on land. Hence, this statement is true.
8	B-False	Note the use of the word "inclusive". This word is used to clarify that Sunday and Friday are also included in the group. But still, Saturday is missing. The statement would have been true if it read "The days between Sunday and Saturday, inclusive, are the days of the week".
9	A-True	Note that the thumb finger and the little finger should not be counted because the statement only uses the word "between".
10	B-False	There are only ten months between January and December excluding January and December.

Answers

FINDING THE TRUTH

Q#	Answer	Reasoning
11	A-True	The word "inclusive" helps us to accurately count January and December as well.
12	A-True	Note the word "Almost". This means that most, but not all birds can fly. For example, an Ostrich is a bird that cannot fly, but most birds do fly.
13	A-True	Elections are won by securing a majority.
14	A-True	Ten inches of anything is longer than five inches of anything.
15	A-True	Eight pounds of sugar is clearly heavier than seven pounds of rice. It does not matter what is being weighed.
16	A-True	Ten gallons of water is obviously more than nine gallons of water.
17	B-False	A big square cannot be smaller than a small square.
18	A-True	This is a true fact that can be verified.
19	B-False	A ten dollar item is more expensive than a seven dollar item.
20	B-False	We can prove that this is false with an example. If the ball is inside the box, then the box cannot be inside the ball.
21	B-False	Tiger and Lion are different types of animals.
22	A-True	This is a true fact. Note the word "attached" used to describe the relationship.
23	A-True	Note the word "part of" used to describe the relationship between earth and solar system and between mountains and earth.

Answers

FINDING THE TRUTH

Q#	Answer	Reasoning
24	A-True	Note the word "part of" used to describe the relationship between the muscles and the bones. Also, note the use of the word "not". Muscles are not part of bones, they are different from bones.
25	A-True	Relationship described here is "belongs to". It is true that snakes and lizards belong to the reptile family.

Answers

DIRECT OR INVERSE RELATIONSHIP

Q#	Answer	Reasoning
1	A-Direct	Obviously, the more one reads, the more words one will learn. Conversely, the less that one reads, the less words one will learn.
2	B-Inverse	The more time we are awake, the less time we are asleep. Conversely, the less time we are awake, the more time we are asleep.
3	A-Direct	The more the number of road accidents, the more the traffic delays. The less the number of road accidents, the less the traffic delays.
4	B-Inverse	The more one spends their money, the less there will be to save. Conversely, the less one spends their money, the more there will be left to save.
5	B-Inverse	The more you play, the less time you will have for studies. The less you play, the more time you will have for studies.
6	B-Inverse	If it rains more, there will be less scarcity of water. Conversely, if it rains less, there will be more scarcity of water. Scarcity means shortage
7	A-Direct	More babies means more population. Less babies means a smaller population.

Answers

DIRECT OR INVERSE RELATIONSHIP

Q#	Answer	Reasoning
8	B-Inverse	The more efficient you are, the less time it will take to do things. The less efficient you are, the more time it will take to do things.
9	A-Direct	The more the number of pages, the longer it will take to read. The less the number of pages, the quicker it will take to read the book.
10	A-Direct	The more the number of vehicles, the more the amount of pollution. If there are less number of vehicles, then the amount of pollution will also be less.

Answers

ANSWERS - CLASSIFICATION

Q#	Answer	Reasoning
1	D	Nail is not part of head; others are.
2	D	Actor is not an elected official; others are.
3	B	Bike has two wheels; others have more than two.
4	B	February starts with an F; others do not.
5	C	Submarine travels inside the water; others do not.
6	A	Air is not a liquid; others are.
7	C	Thief is not a legal occupation; others are.
8	D	Lawn is not a covered space; others are.
9	B	Gram is a measure of weight; others are measures of length.
10	B	Sun is not a planet; others are.
11	B	Left is not an absolute direction like North, South etc.
12	D	Bed is not a kitchen item; others are.
13	A	Swimming is done in water; others are not.
14	D	Flute is played with the mouth; others are not.
15	B	Lion is carnivorous (eats meat); others do not.
16	B	A year is made of months and days. Tomorrow is a relative word.
17	D	A Canadian does not belong to Europe; others do.
18	D	Algebra is not a spoken language; others are.
19	C	Fox is not a domestic animal; others are.
20	B	An Ostrich cannot fly; others can.
21	D	Puddle is a shallow pool of water; others are deeper.
22	D	Magnet is not used for writing; others are.
23	B	Poem is not part of a book; others are.
24	B	Princess is a female; others are male.
25	C	Plate is not meant to hold water; others are.

Answers

WORD ANALOGY

Q#	Answer	Reasoning
1	B	buy : get => sell : give when we buy we get, when we sell, we give
2	B	train : rail => ship : water train travels on rail, ship travels on water
3	B	affection : good => anger : bad affection is a good emotion, anger is a bad emotion
4	B	flower : blossom => tree : grow a flower blossoms and a tree grows
5	B	mouth : cough => nose : sneeze coughing happens through the mouth, sneezing happens through the nose
6	A	rat: rodent => snake : reptile a rat is a rodent, a snake is a reptile; find out what a rodent means
7	A	teeth : grind => hand : squeeze we grind with our teeth, we squeeze with our hand
8	A	radio : sound => television : picture you hear sound from a radio, you see pictures from a television
9	A	ocean : whale => pond : fish you find whales in a ocean; you find fish in a pond
10	B	trash : burn => valuables : keep trash is burnt, valuables are kept
11	B	fruit : juice => corn : oil you make juice from fruit, you make oil from corn

Answers

SENTENCE ANALOGY

1
Mark: I feel like..

What is the analogy?
    Mark's feeling and a deflated balloon

Mark feels
    Answer: B) weak

---

2   Sherry: I can run like a horse.

What is the analogy?
    the way Sherry runs and the way a horse runs

Sherry can run
    Answer: A) fast

---

3   Brian: Shelly sings like a cuckoo.

What is Brian comparing?
    the way Shelly sings and the way a cuckoo sings

According to Brian, Shelly's voice is
Answer: A) nice to hear.
       A cuckoo is a bird whose voice is considered very musical.

Answers
© Gift Of Logic, Inc * Copying prohibited

SENTENCE ANALOGY

4 John held..

What is the comparison?
  John's powerful grip and a crocodile's powerful bite

John held Mike's hand
  Answer: B) tightly

---

5 Maggie: Susan won a gold medal...

What is Pam comparing?
  Susan's achievement and music

Pam says that the news of Susan winning a gold medal is
  Answer: B) good news to her

---

6 The news spread like wildfire.

What is being compared?
  news and wildfire

Which one of the following is true?
Answer: A) news spread fast

Answers
© Gift Of Logic, Inc * Copying prohibited

AGREE-DISAGREE

1

Jack: Apple is the tastiest ..   Jill: An Orange tastes better..
Answer: B) disagree with each other that apple is the tastiest fruit.

Reasoning: Jack indicates that apple is the tastiest fruit. But, Jill indicates that an Orange tastes better than an apple. So, they disagree with each other that apple is the tastiest fruit.

---

2   Mary: .. book once.
Mark: .. book more than once.

Mary and Mark Answer: B) disagree on the number of times the book must be read.

Reasoning: Mary says that the book must be read once. So, you could read it just one time and satisfy her. Mark says that the book must be read more than once. So, according to him, reading the book just one time is not sufficient.

Therefore, they are in disagreement on the number of times the book must be read.

Answers

© Gift Of Logic, Inc * Copying prohibited

AGREE-DISAGREE

3 Walid: We should not..   Mohan: We should..
Answer: A) agree with each other.
Reasoning: To be tardy is to be late. Walid says that we should not be late, which is the same as saying that we should be punctual. This is what Mohan also says.

---

4 Mia: ... to learn.   Lee: ... to make friends
Answer: B) disagree with each other on the purpose of going to school.
Reasoning: Mia thinks that the purpose of going to school is to learn while Lee thinks that the purpose of going to school is to make friends.

Answer: A) agree with each other that every child must go to school.
Reasoning: Both agree that every child must go to school.

---

5   Raja: We must sleep..   Rashid: We must work..
Answer: B) have different opinions on how to remain healthy.

Reasoning: Raja says that we must sleep a lot to remain healthy whereas Rashid says that we must work a lot to remain healthy. These are clearly different opinions on how to remain healthy.

---

6 Sonia: We can use as much .. Maggie: We must be frugal ..
Answer: B) disagree on how to use electricity.

Reasoning: Sonia says that we can use as much electricity as we want, but Maggie says that we must be frugal in using electricity. To be frugal is to avoid waste. So, they differ in their opinions.

Answers

© Gift Of Logic, Inc * Copying prohibited

INFERENCE - must be true

1 A country is divided into... Answer: A) It belongs to a state and a country.
Reasoning: Reading the two statements carefully will help you answer the question correctly. Drawing a picture of a country, divided into states and cities will help you make the inference. A city belongs to a state and a state belongs to a country.

2 Every car driver ... Answer: B) If a man has a driver's license, he will have an address.
Reasoning: The statement says that the driver's license shows the driver's name and address. So, we can infer that if a man has a driver's license, he will have an address. Choice A is incorrect because if a man has an address, it does not mean that he should have a driver's license. He may have an address, but may not drive a car at all.

3 Except first and third grade... Answer: B) A second grade and a fourth grade student can go to the museum.
Reasoning: Note carefully the use of the word "except". The statement means that, anyone in the first grade and the third grade cannot go to the museum. Choice A is incorrect, because it has a first grader in it. Choice B is correct - a second grade and a fourth grade student can go to the museum.

4 Recycling truck...
Answer: B) The recycling truck will not come on Sundays.
Reasoning: Sunday being a weekend day, the truck will not come.

Answers
© Gift Of Logic, Inc * Copying prohibited

INFERENCE - must be true

5 Botany is a branch...
Answer: B) Oak trees are studied in Botany. <u>Reasoning:</u> The given statement says that Plants are studied in Botany. A Lion is not a plant. So, Lions are not studied in Botany. So, choice A is incorrect. Oak tree is a tree and trees are studied in Botany. So, we can infer that Oak trees are studied in Botany.

---

6 The Green Leaf party... Answer: B) Baker is the winner.
<u>Reasoning:</u> The Blue Leaf party won the election. Baker belongs to the Blue Leaf party. So, we can infer that Baker won the election.

---

7 Residents of ... Answer: A) John dug at least two hundred feet.
<u>Reasoning:</u> From the given facts, gold can be found after digging to a depth of 200 feet. This means that one has to dig at least 200 feet before finding gold.

---

8 The population of a city ... Answer: B) More people lived in the city in 2000 than in 1990.
<u>Reasoning:</u> The population increased from 1990 to 1995 and then remained flat, meaning it did not increase from 1995 to 2000. So, the same number of people lived in 1995 and 2000. But, since there was an increase from 1990 to 1995, more number of people lived in 2000 than in 1990.

---

9 The new Liberty High School can have ... Answer: B) More than 300 students will attend the school.
<u>Reasoning:</u> The maximum allowed is 350 students and this limit was reached. So, we can infer that more than 300 students will attend the school. Choice A is clearly incorrect.

Answers
© **Gift Of Logic, Inc** * **Copying prohibited**

INFERENCE - must be true

10 Physical education classes ... Answer: A) Physical education classes are held on the same days every week.

Reasoning: The confusion here is between days and weeks. The statement says that classes are held on Mondays and Fridays every week. So, we can infer that the classes are held on the same days every week. If choice B were to be true, then the classes would have to be held on different days every week, which is not true.

INFERENCE - cannot be true

11 An emergency ... Answer: A). The fire was put out at 8 PM.

Reasoning: Note that the **correct answer is the one that cannot be true.** Pay attention to timing and the word "after" in the answer choice. The fire station received the call at 9 PM. So, obviously it is not possible (that is, cannot be true) for the fire in the building to have been put out at 8 PM.

12 The Creek Park... Answer: B) Five teams can play at the same time.

Reasoning: The **correct answer is the one that cannot be true.** Since each game requires two teams to play, the following number of teams can play at any given time in the five soccer fields: 2 teams, 4 teams, 6 teams, 8 teams, 10 teams. So, five teams cannot play at the same time as one of them will be left without a team to play against. Choice A can be true - five games can be played at the same time since there are five fields.

Answers
© Gift Of Logic, Inc * Copying prohibited

INFERENCE - cannot be true

13 Jason was the first in line... Answer: A) John, who was also in the line, was the first one to purchase the novel.
Reasoning: Since Jason was the first in line, and since books are sold on a first come-first served basis, it can be inferred that Jason would be the first one to buy the popular novel. This answer choice says that John was the first one to buy the popular novel. This cannot be true and so, it is the correct answer.

14 Two cars, one red and one blue .... Answer: B) The blue car won the race. Reasoning: From the facts, it is clear that the red car moved faster than the blue car. Hence it cannot be possible for the blue car to win the race. The choice that cannot be true is the correct answer.

INFERENCE - True or False

1 The following table... Answer: B) False.
Reasoning: Ray will wear green pants since he is six and a half years old.

2 A road goes ... Answer: B) False.
Reasoning: Since the cement truck is taller than the tunnel, it cannot go through the tunnel.

3 A magnet ... Answer: A) True.
Reasoning: The facts state that the opposite poles of two magnets attract each other. That is, the north pole attracts the south pole. The red magnet stays attached to the blue magnet, meaning that the poles that connect them are of different types. We know that the south pole of the red magnet touches the blue magnet. So, only the north pole of the blue magnet can keep them attached together.

INFERENCE - True or False

4 Jean is six feet tall ...

Answer: B) False.

Reasoning: It is true that Jean is taller than Joan. But, Joan is not shorter than Jane. Joan is five feet tall and Jane is four feet tall. So, Joan is taller than Jane. But, the statement says that Joan is shorter than Jane, which is false.

Note that the statement uses the word "and" to describe two facts, one of which is true and the other is false. Both the facts have to be true for the entire statement to be true.

---

5 Circle A and circle B.. Answer: B) False.

Reasoning: Drawing the circles will help you visualize the stated facts and draw inferences. Just because circle A and circle B are inside circle C does not mean that circle A must be completely inside circle B. While that is possible, we cannot make that conclusion as there are other possibilities as well. For example, it is also possible that only a portion of circle A is inside circle B or circle A and circle B are totally outside each other. The diagram below shows these two possibilities.

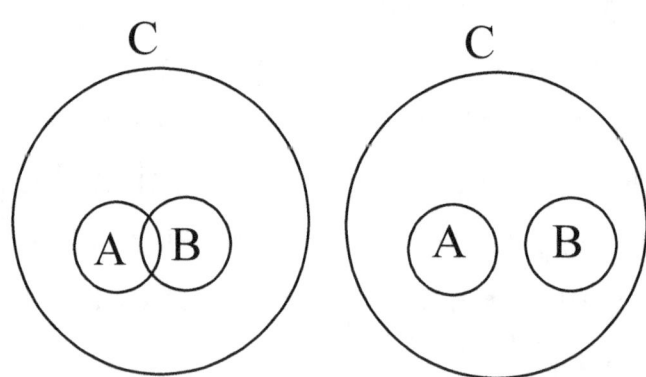

Answers

SEQUENCING

1

* Write the days of the week in sequence starting from Sunday.
Answer: Sunday, Monday, Tuesday, Wednesday, Thursday, Friday, Saturday.

* Write the months of the year, in sequence, starting from the first month.
Answer: January, February, March, April, May, June, July, August, September, October, November, December.

* Write all the objects in a table that you can see, in ascending order.
Answer: Notebook, Pen, Pencil, Scale (your answer could be different)

* Write the name of all members of your family in descending order of their first names.
Answer: Zac, Walton, Aaron (your answer could be different)

* Write the names of few cities in your state in descending order.
Answer: San Antonio, Houston, Dallas, Austin (your answer could be different)

* Write all the places you visited yesterday in sequence from morning to evening.
Answer: School, Post Office, Mall (your answer could be different)

Answers
© Gift Of Logic, Inc * Copying prohibited

SEQUENCING

2
Three movies will..

1) When will each of the three movies begin?
   Answer: 4 PM, 7 PM and 10 PM

2) At what time will the second movie end?
   Answer: 9 PM

3) At what time will the last movie end?
   Answer: 12 AM

3
A theater complex..

Theater A's movie timings are: 4 PM, 6 PM, and 8 PM
Theater B's movie timings are: 5 PM, 7 PM, and 9 PM

1) At what time will the fourth movie in the complex be screened?
   Answer: 7 PM

2) In which theater will the second movie in the complex be screened?
   Theater B at 5 PM

3) When and where will the last movie in the complex be screened?
   9 PM at theater B

Answers
© Gift Of Logic, Inc * Copying prohibited

SEQUENCING

4

A sequence of red..

Row#1: Red, Green, Blue
Row#2: Yellow, Blue, Orange

1) At any particular time, how many lights flash?
Answer: A) 2. At any given time, one light in each row would be flashing.

2) When the blue light in Row#2 has finished flashing, how many lights would have finished flashing
- in Row# 1?   Answer: 2 (red and green)
- in Row# 2?   Answer: 2 (yellow and blue)
- in Total?   Answer: 4

3) When the blue light in row#1 flashes, which one of the following lights in row#2 also will flash?

Answer: C) Orange. Blue light is the third light in row#1. Orange light is the third light in row#2. The lights in the two rows flash at the same time.

4) Before the orange light starts flashing, how many lights would have finished flashing in total?
    Answer: B) 4. Before the orange light starts flashing, the red and green lights in row# 1 and the yellow and blue lights in row# 2 would have finished flashing. Note that the orange light in row#2 will start flashing at the same time as the blue light in row#1.

Answers
© Gift Of Logic, Inc * Copying prohibited

SEQUENCING

5 A loud drum beat..

1) Represent the sequence in symbols.

    ! ~ ~ ! ~ ~ ! ~ ~ ! ~ ~ !   ~ ~ ! ~ ~   total of 6 sequences
    1 2 3 4 5 6 7 8 9 10 11 12 13 14 15 16 17 18

Note that "repeating" means "do something over again". If a sequence of drum beats is repeated five times, then including the first drum beat, we will have a total of 6 drum beats.

2) The fifth drum beat was loud.
Answer: B) False.
Reasoning: The fifth drum beat is soft, as shown in the sequence by the ~ symbol.

3) Every even drum beat was soft and every odd drumbeat was loud.
Answer: B) False. Some even drum beats were loud (drum beats 4,10,16) and there were odd drum beats that were soft (drum beats 3,5,9,11,15,17)

---

6 Several ants marched.. Answer: C) 6.

Reasoning: Starting from the second ant, every other ant is female. The third female ant is the sixth ant in the sequence.

1	2	3	4	5	6	7	..so on
	F		F		F		

F represents a female ant.

Answers
© Gift Of Logic, Inc * Copying prohibited

SEQUENCING

7 The following chart shows the sequence..

1) How many science classes start before noon on Monday?
   Answer: 2 (at 10 AM and 11 AM)
2) How many math classes start after 11 AM?
   Answer: 1 (on Thursday)
3) How many science classes are held before Wednesday?
   Answer: 3 (2 on Monday and 1 on Tuesday)
4) How many Reading classes start before 11 AM Thursday?
   Answer: 4 (1 on Monday, 2 on Tuesday, 1 on Wednesday)
5) Write the sequence of classes held at noon.
   Answer: Reading, Science, Science, Math, Reading

8 The following is eye doctor..

1) How many patients will the doctor see today? Answer: 6 (the break is not to be counted as a patient)

2) The first appointment for Dr. Frank is before 9 AM. Answer: B) False. The first appointment is not before 9 AM, but at 9 AM.

3) Mary will be seen by the doctor before Mark is seen. Answer: A) True. Mary will be seen at 10:30 AM and Mark will be seen at 2 PM.

4) Laura and Roger will see the doctor before Steve. Answer: B) False. Roger will see the doctor only after Steve.

5) Laura and Steve will see the doctor after Mary and Ron.
 Answer: B) False. Laura and Steve will see the doctor before Mary and Ron.

Answers
© Gift Of Logic, Inc * Copying prohibited

SEQUENCING

9 Seven cars tagged as C1,C2,C3,C4,C5,C6, and C7..

1) If the positions of cars C3 and C4 are interchanged, then the positions of cars before C3 and cars after C4 will change.
Answer: B) False Reasoning: Since the cars C3 and C4 are interchanged, only their positions will change, but not the positions of other cars.

2) If car C3 is removed from the line-up, then the fifth car in the line up will be Answer: B) C6 Reasoning: Since car C3 is removed from the lineup, the cars currently in the line-up are C1,C2,C4,C5,C6,C7. So, the fifth car is C6.

---

10   When Joe was.. Soap, Popcorn, Juice, Shoes ..

1) What is the first food related advertisement that he saw?
    Answer: Popcorn
2) What is the last food related advertisement that he saw?
    Answer: Juice
3) Which non-food related advertisement as shown after the advertisement for popcorn?
    Answer: Shoes
4) Which non-food related advertisement was shown before the advertisement for shoes?
    Answer: Soap
5) Which advertisement did Joe see between the Soap and Juice advertisements?
    Answer: Popcorn

Answers
© Gift Of Logic, Inc * Copying prohibited

SEQUENCING

11 Sam picked up ..
The sequence of the carts that were brought into the store in ascending order is Answer: A) 2, 3, 6, 9

12 There were several chairs..
Answer: C) 5, 3, 2, 1. Stephanie stacked the chairs with the lowest number at the bottom. This means the higher numbered chair will be at the top.

13 Symbol $\boxed{\rightarrow 1}$ means..
1) If the instructions in the symbols shown above are followed, then what does the following sequence of symbols mean taken together?

$\boxed{\rightarrow 1}\boxed{\leftarrow 1}$

Answer: D) no change in position. The symbols cancel each other out.

2) $\boxed{\rightarrow 2}$ and $\boxed{\leftarrow 1}$ effectively mean the same.
Answer: A) True
$\boxed{\rightarrow 2}$ means 2 steps forward and 1 step backward - effectively 1 step forward. $\boxed{\leftarrow 1}$ means 1 step backward and 2 steps forward, effectively 1 step forward. So, they mean the same.

3) Draw a sequence of symbols that will take someone five steps forward.
$\boxed{\rightarrow 2}\boxed{\rightarrow 2}\boxed{\rightarrow 2}\boxed{\rightarrow 2}\boxed{\rightarrow 2}$

The above set of symbols is one way to do this. Each of the symbols is effectively one step forward. So, five of these symbols will take you five steps forward. Think of other ways to take five steps forward.

Answers
© Gift Of Logic, Inc * Copying prohibited

SEQUENCING

14

Construction project plans for two..

1) If construction for both houses begins on the same day, then the construction for both houses will be completed on the same day.

Answer: B) False

Reasoning: House# 2 will take ten fewer days to construct than house# 1. So, if the construction for both houses began on the same day, the construction of both houses will not be completed on the same day. House# 2 will be completed sooner than house #1.

2) If construction for both houses begins on the same day, then the tile laying task for both houses will be completed on the same day.

Answer: A) True
Reasoning: It would take 100 days from the start of construction for the tiles task to be completed in both houses.

3) Painting for house#2 will start before painting for house# 1.
Answer: A) True

Reasoning: Painting for house# 1 will begin after 100 days. Painting for house# 2 will begin after 90 days. So, the painting for house#2 will start before the painting for house# 1.

Answers
© Gift Of Logic, Inc * Copying prohibited

SEQUENCING

15

Three boats with two people each in them participated in a boating competition..

1) The first boat reached the finish line first.
Answer: B) False.
Reasoning: The second boat reached the finish line at 11 AM, before any other boat. The first boat reached the finish line at noon, which is after 11 AM. Noon is 12 PM.

2) Which of the following must be true?
Answer: A) The boat that takes the least time to reach the finish line is the winner.
Reasoning: The boats started the race at different times. So, we need to find how long it took for them to reach the finish line and decide the winner. We cannot just assume that the boat that arrived first at the finish line is the winner. This would be true if they all started at the same time, but this is not the case.

3) The order of the boats in terms of how long they took to reach the finish line, from least to most is
Answer: C) boat#2, boat#3, boat#1

Reasoning: Boat#2 took 2 hours (11 AM-9 AM=2 hours), boat#3 took 3 hours (1 PM- 10 AM=3 hours) and boat#1 took 4 hours (12 PM-8 AM=4 hours) in total to reach the finish line from the starting line.

SCHEDULING

1
A cardiologist sees..

1) The cardiologist did not see patients on Saturdays.
Answer: B) False.
Reasoning: He begins seeing patients from September 1st, which happens to be a Saturday. He also saw patients on the 15th and 29th, which are Saturdays.

2) The cardiologist did not see patients on Sundays.
Answer: B) False.
Reasoning: He saw patients on the 9th and 23rd, which are Sundays.

2
After an accident ..
   Josh, Jamie, Jane is the order in which they arrived at the hospital.

1) Jamie will be scheduled to be seen before Josh.
Answer: B) False.
Reasoning: Josh arrived at the hospital first and must be scheduled before Jamie.

2) Jamie will be scheduled to be seen before Jane.
Answer: A) True
Reasoning: Jane was the last one to arrive and hence she will be the last one to be scheduled. Since Jamie also arrived at the hospital and she is not the last one, she will be scheduled before Jane.

Answers
© Gift Of Logic, Inc * Copying prohibited

SCHEDULING

3

A hospital has ..

Which of the following schedules of the three doctors is correct?

Monday	Tuesday	Wednesday	Thursday	Friday	Correct?
A	B	A	C	C	correct
B	B	A	C	C	correct
A	B	C	A	B	incorrect

Reasoning: Apply the schedules of the three doctors against the schedule shown and spot the error. The third schedule is incorrect because it conflicts with the schedules of all the three doctors.

4

Doctor Even works on even numbered days.
Doctor Odd works on odd numbered days.

1) Since doctors Even and Odd take alternate turns to work, they work the same number of days each month.
Answer: B) False.
Reasoning: On months with 31 days, doctor Odd will work more number of days than doctor Even.

2) Doctor Even works on the last day of April, June, September and November. Answer: A) True
Reasoning: Since the last day of these months is an even numbered day, doctor Even will be working on these days.

Answers

© Gift Of Logic, Inc * Copying prohibited

STRATEGY PROBLEMS

1 There are seven people..

Answer: A) move 2 people form team A to team B.

<u>Reasoning</u>: Clearly, this strategy will make both teams have five people.

---

2 There are 4 boys and 2 girls..

Answer: B) move 1 boy from team Orange to team Apple and 1 girl from team Apple to team Orange

<u>Reasoning:</u> This question seems to be confusing, but you can sort things out by charting out the facts in a table as shown. The start row shows the number of boys and girls in each team. The other three rows show the movement of boys from one team to another for each of the choices. When one boy is transferred from team Orange to team Apple, team Orange loses one boy and team Apple gains one boy. We mark this fact with -1 boy in team Orange and + 1 boy in team Apple. In the second row for each choice, we total up the number of boys in the start row and the number of boys in the choice row. As you can see, choice B is the correct strategy to have the same number of boys and girls in each team.

	Team Orange		Team Apple	
Start	4 boys	2 girls	2 boys	4 girls
choice A	-1 boy	-1 girl	+ 1 boy	+ 1 girl
	3 boys	1 girl	3 boys	5 girls
choice B	-1 boy	+ 1 girl	+ 1 boy	-1 girl
	3 boys	3 girls	3 boys	3 girls
choice C	+ 1 boy	-1 girl	-1 boy	+ 1 girl
	5 boys	1 girl	1 boy	5 girls

Answers

STRATEGY PROBLEMS

3 Gardener Graham..

First, chart out the facts so that it becomes easy to find the solution.

	Garden-A	Garden-B
start:	A B C	D E F
actual:	~~A~~ B C	D ~~E~~ F
also:	B ~~E~~	~~A~~ D

Now, represent the plants in garden A and B for each choice.

choice A:	E B C	A D F
choice B:	E F B	A C D
choice C:	E D C	A B F

Answer: C

Reasoning: Choice A is incorrect. If we move plant A from garden A to garden B, and plant E from garden B to garden A, then we are left with plants E, B, and C in garden A and plants A, D, and F in garden B. But, plants B and E will not grow in garden A and plants A and D will not grow in garden B. So, this strategy will not make all plants grow.

Choice B is also incorrect since it will leave us with E, F, and B in garden A and A, C, and D in garden B. B and E cannot be together in garden A.

Choice C is the correct strategy as this leaves us with E, D, and C in garden A and A, B, and F in garden B. This strategy will allow all the plants to grow.

Answers

STRATEGY PROBLEMS

4 Jenny started her day..

Answer: C) do homeworks 1 and 3 before noon and then, do homework-2.

Reasoning: There are two homeworks of 2 hours duration each, and one homework of 4 hour duration. Jenny has only 4 hours in the morning after 8 AM and must take a break at noon.

Choice A cannot be correct because homeworks 1 and 2 will take up 6 hours in the morning, but she has only 4 hours available before noon.

Choice B is incorrect as well because homeworks 2 and 3 together will take 6 hours to do and this cannot be done before noon when she has to take a break.

Choice C is correct because homeworks 1 and 3 can be done in the morning for 4 hours, then she can take a break for 1 hour, and then do homework-2 for four hours.

---

5 Lee is planing to visit..

Answer: A) go to his sister's house first, and then to his uncle's.

Reasoning: If he went to his sister's house first, it will take 30+ 20+ 20 minutes. If he went to his uncle's house first, it will take 20+ 40+ 30 minutes. Choice A will take the shortest time to visit his sister and uncle. Drawing a picture with the distances will help to visualize this problem.

Answers

STRATEGY PROBLEMS

6   There are four liquids.. B and C when mixed, will cause a loud noise. A and D when mixed, will cause fire.

Answer: A) mix A, D, and B.

<u>Reasoning</u> If A, D, and B are mixed, there will be fire because of A and D mixing together. But, there will not be a loud noise because B and C are not there in this mix.

Is there any other 3-solution compound that will not cause a loud noise?
   A, D, and C, when mixed together will not cause a loud noise.

---

7   Vaccine-X can kill germs A and B, but not C.
Vaccine-Y can kill germs B and C, but not A.
Vaccine-Z can kill germs C and A, but not B.

Answer: B) give any two vaccines.

<u>Reasoning</u> Any two vaccines will kill all the three germs A, B ,and C. One vaccine alone will not kill all the three germs. All the three vaccines can kill all the three germs, but we need to find a solution that uses the least number of vaccines. So, choice B is the correct answer.

Answers
© Gift Of Logic, Inc * Copying prohibited

STRATEGY PROBLEMS

8 At 4 PM, the pilots..

The best strategy for the pilot is to
Answer: B) land the airplane in bad weather.

Reasoning: It is wise to land the airplane before the fuel runs out. Since the fuel will last for only 4 hours whereas the bad weather is expected for 5 hours, the correct strategy would be to land the airplane in bad weather.

---

9 Jack can take either a 3-hour ..
Answer: B) take the 4-hour flight.

Reasoning: Taking the 4-hour flight will land Jack at Denver at 5 PM, giving him a 1-hour stopover at Denver. Taking the three hour flight will land him at Denver at 4 PM, giving him a 2-hour stopover. So, to have the least stopover, it is better to take the longer flight.

---

10 Ferry#1 and Ferry#2 carry..
Answer: B) take Ferry#1 and Ferry #2.

Reasoning: Choice A is incorrect, because Ferry#1 leaves island B at 9 AM, when he will be in a meeting there. So, he cannot travel on Ferry#1 to island D for his 11:45 AM meeting. Choice B is correct, because Martin can take Ferry#1 to island B, arrive at island B at 8:55 AM to attend the 9 AM-9:15 AM meeting, and then after the meeting, take Ferry#2 at 9:30 AM to island D, reaching island D at 11:25 AM for his 11:45 AM meeting.

Answers
© Gift Of Logic, Inc * Copying prohibited

STRATEGY PROBLEMS

11

The Mayor of Clean City..

Assuming that all the trucks are of the same size, the best strategy for the City manager to remove the garbage quickly is

Answer: A) to use maximum number of trucks possible to haul the garbage.

Reasoning: This is obvious. We don't know how many trucks are available to haul the garbage, but since we have to pick between the maximum number of trucks and a minimum number, it is obvious that using the maximum number of trucks will help remove the garbage quickly.

Assuming that all the trucks are of the same size, the best strategy for the City manager to remove the garbage with the least expense is

Answer: B) to use the minimum number of trucks to haul the garbage.

Reasoning: This is clear since using less number of trucks means using less fuel, workers, etc which in turn will keep the expenses to the minimum.

STRATEGY PROBLEMS

12 Ali was going to..
Answer: B) take the detour.

Reasoning: The detour is only 18 miles and he has enough fuel for 20 miles. So, he can reach the office without running out of fuel.

13
Goldstein has to deliver furniture...

1) To deliver the furniture quickly, Goldstein should take
Answer: B) the 60 mile road.

Reasoning: This road only takes 2 hours and is the quickest.

2) Goldstein gets rewarded for driving the shortest distance. To get the maximum reward, Goldstein must take

Answer: A) the 40 mile road.

Reasoning: This road is the shorter than the 60 mile road. Hence, driving on this road will get him the reward.

Answers
© Gift Of Logic, Inc * Copying prohibited

1 POSITIONING PROBLEMS - Vacancy

1) After all the butterflies sit on the branch, how many spots will be without a butterfly? Answer: A) 1

Reasoning: Clearly, there are three butterflies and four spots. So, one spot will be vacant without a butterfly.

2) If two butterflies sit in spots 1 and 3 respectively, then which of the following spots will be vacant?   B) 2 or 4, but not both

Reasoning: If spots 1 and 3 are taken, then two butterflies would have sat in those positions. This leaves the third butterfly to sit in either spot 2 or 4. If the third butterfly sits in spot 2, then spot 4 will be vacant. If it sits in spot 4, then spot 2 will be vacant. So, choice B is the correct answer. Choice A is incorrect since both the spots cannot be vacant.

3) Regardless of where the three butterflies sit, there will always be one vacant spot. Answer: A) True

Reasoning: The three butterflies can sit in any of the three spots, thereby leaving one spot always vacant. There are more spots than butterflies.

Answers

© Gift Of Logic, Inc * Copying prohibited

2 POSITIONING PROBLEMS - Vacancy

1) If all the three butterflies must sit next to each other, then which of the following spots can be vacant? Answer: A) Spots 1 or 4

Reasoning: If the three butterflies sit in spots 1,2, and 3, then spot 4 will be vacant. If all the butterflies sit in spots 2,3, and 4, then spot 1 will be vacant. Therefore choice A is the correct answer. Choice B (2 or 3) is incorrect because if spots 2 or 3 are vacant, then all the three birds cannot sit next to each other.

2) If no butterfly can sit in spot 3, then which one of the following must be true? Answer: B) There will be two butterflies to the left of spot 3.

Reasoning: Since no butterfly can sit in spot 3, this means that two butterflies will sit in spots 1 and 2 and one butterfly in spot 4. So, there will be two butterflies to the left of spot 3.

Answers
© Gift Of Logic, Inc * Copying prohibited

3   POSITIONING PROBLEMS - Vacancy

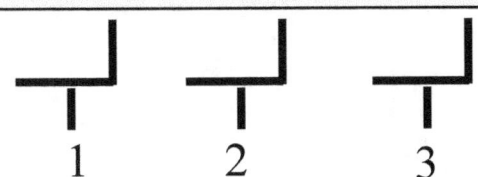

1) If Ying must sit to the left of Zac, then which of the following seating arrangements is possible? Answer: A, B, C - all the seating arrangements are possible.

Reasoning: To answer this question, write down the rule in short form. The rule is that Ying must sit to the left of Zac. Write this rule as Ying-Zac and look for the Ying-Zac pattern of seating. Note that there can be a vacant spot between Ying and Zac as long as Ying is to the left of Zac.

1)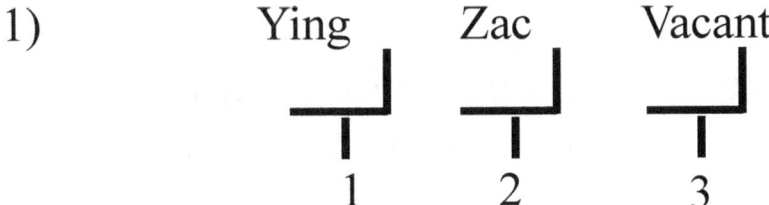

2)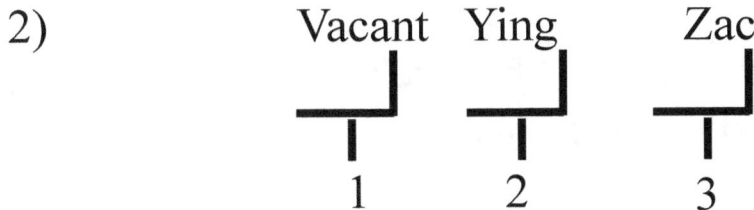

3)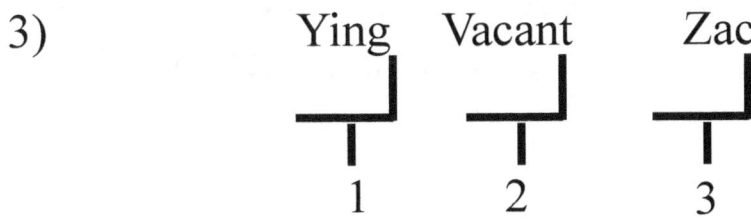

Answers
© Gift Of Logic, Inc * Copying prohibited

| 4 | POSITIONING PROBLEMS - No vacancy |

π   π   π
1   2   3

1) How may students will be left without a stool?
Answer: A) 1

Reasoning: There are four girls, but only three stools. So, one girl will be left out without a stool.

2) If A and B are seated, then who will not be able to get a stool?
    Answer: B) C

Reasoning: The rule is "If A gets a stool, then D must also get a stool". The question says that A and B are seated, which means A, D and B are seated. So, we can infer that C is not seated.

3) Which one of the following statements must be true?
Answer: B) D, B, and C can be seated in the stools.

Reasoning: The rule says that if A is seated, D must also be seated. In this choice, A is not seated. It is ok for D to be seated without A being seated. So, this choice is correct. Choice A is incorrect since A is seated, but D is not, thus violating the rule.

Answers
© Gift Of Logic, Inc * Copying prohibited

5   POSITIONING PROBLEMS - No vacancy

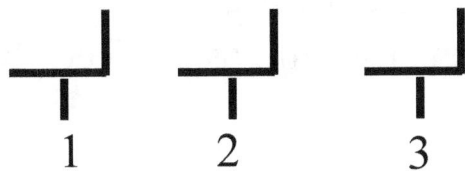

The rules are: If A then C, and If B then D.

1) How may students will be left without a chair?
Answer: B) 2
<u>Reasoning:</u> There are five students, but only three chairs. So, two students will be left without a chair.

2) If A and D are seated, then who will not be able to get a seat?
Answer: A) B and E
<u>Reasoning:</u> If A gets a seat then C must get a seat. D also gets a seat, so A, C and D are seated. This means that B and E are left without a seat.

3) Which one of the following statements must be true?
Answer: B) C, D, and E can be seated in the chairs.
<u>Reasoning:</u> The rule says "If B then D". The seating arrangement C, B, and E does not have D in it. So, choice A is incorrect.

Choice B is correct. The seating arrangement, C, D, and E does not violate the rules. Note that the rule "If A then C" is not violated by having C seated, but not having A seated. It is okay to seat C but not seat A, but it is NOT okay to seat A, but not seat C. Similarly, it is okay to seat D and not seat B. The "If A then C", and "If B then D" rules are called conditional rules. You will learn about them in detail in workbook# 3.

Answers
© Gift Of Logic, Inc * Copying prohibited

| **1** | PICKING PROBLEMS |

If Red is picked, then Green must be picked.
  >> You can write and remember this rule as "if Red then Green" or "if R then G".
If Green is picked, Blue must be picked.
>> You can write and remember this rule as "if Green then Blue" or "if G then B".

1) Which of the following picks is correct?
Answer: B) Green, Blue, White
<u>Reasoning</u>: This selection satisfies the rule "if G then B". It does not violate the rule "if R then G" because a Red ball is not selected in this choice.

2) If Red is picked, then White and Pink will not be picked.
Answer: A) True
<u>Reasoning</u>: If R then G, and If G then B, means if Red is picked then Green and Blue will be picked. Since only three balls must be selected, this means that white and pink cannot be selected.

3) White and Pink cannot be picked together.
Answer: B) False
<u>Reasoning</u>: White, Pink, and Blue can be selected together.

4) How many balls will always not be picked?
Answer: A) 2. Out of five balls, if you can select only three, two balls will be always left out from selection.

Answers 130
© Gift Of Logic, Inc * Copying prohibited

2	PICKING PROBLEMS

Two boxes contain balls of different colors as shown. A total of five balls must be selected from these two boxes.

If one red ball is selected, two green balls must also be selected.
   >> if one red then two green - remember this rule as - if 1R then 2G

If one green ball is selected, then two blue balls must be selected.
   >> if one green then two blue - remember this rule as - if 1G then 2B

1) Which of the following statements must be true?
Answer: A) A Red ball cannot be selected.
Reasoning: if 1R, then 2G
             if 1G then 2B
             so, if 1R then 2B, 2B (for each of the two G's)
But, there are only two Blue balls, whereas we need four. So, the Red ball cannot be selected.
Choice B - A green ball can be selected without a blue ball - cannot be true since it violates the "if 1G then 2B" rule.

2) Which one of the following selections is correct?
Answer: A) Blue, Blue, White, Pink, Green
Reasoning: Since the rule is "if 1G then 2B", this selection is correct. Since there is no Red ball in this selection, the other rule " if 1R then 2G" does not apply.

Choice B) "Green, Blue, White, Pink, Green" is incorrect because it violates the rule "if 1G then 2B" as the selection does not have two blue balls.

Answers
© Gift Of Logic, Inc * Copying prohibited

NUMERIC SUDOKU

1

1		2	
	2		
4		3	
		1	

1	4	2	3
3	2	4	1
4	1	3	2
2	3	1	4

2

4			3
		4	
3		1	
			2

4	1	2	3
2	3	4	1
3	2	1	4
1	4	3	2

NUMERIC SUDOKU

3

		2	
1			3
	4	1	

4	3	2	1
1	2	4	3
2	1	3	4
3	4	1	2

4

4			
		4	2
	4	3	
3	1		

4	2	1	3
1	3	4	2
2	4	3	1
3	1	2	4

Answers
© Gift Of Logic, Inc * Copying prohibited

NUMERIC SUDOKU

5

2			3
3			
	2		4
		1	

2	1	4	3
3	4	2	1
1	2	3	4
4	3	1	2

6

	3		1
2		4	
	2		
1			2

4	3	2	1
2	1	4	3
3	2	1	4
1	4	3	2

NUMERIC SUDOKU

7

	3	2	
	1		
4		3	

1	3	2	4
2	4	1	3
3	1	4	2
4	2	3	1

8

3	4		
		3	
1	3	4	2

3	4	2	1
2	1	3	4
4	2	1	3
1	3	4	2

Answers 135
© Gift Of Logic, Inc * Copying prohibited

NUMERIC SUDOKU

9

	1	2	
			4
	2		1

4	1	2	3
2	3	1	4
1	4	3	2
3	2	4	1

10

1			4
4		1	
		4	
3		2	

1	2	3	4
4	3	1	2
2	1	4	3
3	4	2	1

Answers

NUMERIC SUDOKU

11

2			3
		2	1
	2		4
3			

2	1	4	3
4	3	2	1
1	2	3	4
3	4	1	2

12

1			3
		2	
3		1	
	1		4

1	2	4	3
4	3	2	1
3	4	1	2
2	1	3	4

Answers
© Gift Of Logic, Inc * Copying prohibited

ALPHABETIC SUDOKU

1

A			
	B		A
D		C	
		A	

A	D	B	C
C	B	D	A
D	A	C	B
B	C	A	D

2

D			
B		D	
C		A	
			B

D	A	B	C
B	C	D	A
C	B	A	D
A	D	C	B

Answers 138
© Gift Of Logic, Inc * Copying prohibited

ALPHABETIC SUDOKU

3

		B	
A			C
	D		

D	C	B	A
A	B	D	C
B	A	C	D
C	D	A	B

4

D			
		D	B
	D	C	
	A		

D	B	A	C
A	C	D	B
B	D	C	A
C	A	B	D

Answers 139

ALPHABETIC SUDOKU

5

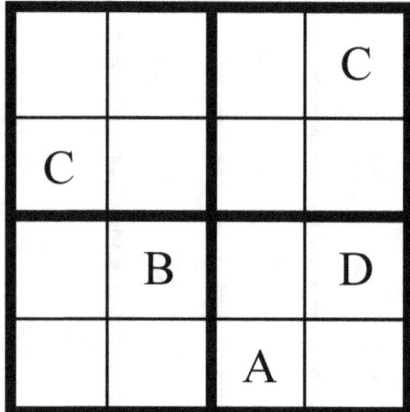

B	A	D	C
C	D	B	A
A	B	C	D
D	C	A	B

6

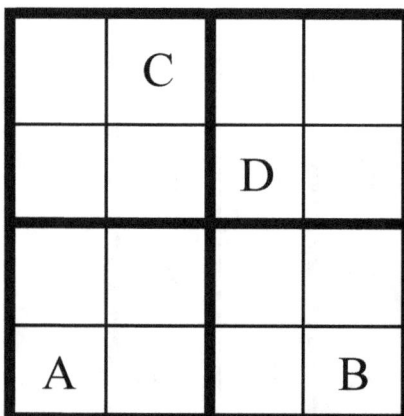

D	C	B	A
B	A	D	C
C	B	A	D
A	D	C	B

Answers

ALPHABETIC SUDOKU

7

	C	B	
	A		
D		C	

A	C	B	D
B	D	A	C
C	A	D	B
D	B	C	A

8

	D		
		C	
		A	
A	C		

C	D	B	A
B	A	C	D
D	B	A	C
A	C	D	B

Answers 141

© Gift Of Logic, Inc * Copying prohibited

ALPHABETIC SUDOKU

9

	A	B	
			D
	B		A

D	A	B	C
B	C	A	D
A	D	C	B
C	B	D	A

10

A			D
D			
B		D	
		B	

A	B	C	D
D	C	A	B
B	A	D	C
C	D	B	A

Answers
© Gift Of Logic, Inc * Copying prohibited

ALPHABETIC SUDOKU

11

B			
		B	A
	B		D
C			

B	A	D	C
D	C	B	A
A	B	C	D
C	D	A	B

12

A			C
C		A	
	A		D

A	B	D	C
D	C	B	A
C	D	A	B
B	A	C	D

Answers

PICTURE SEQUENCE

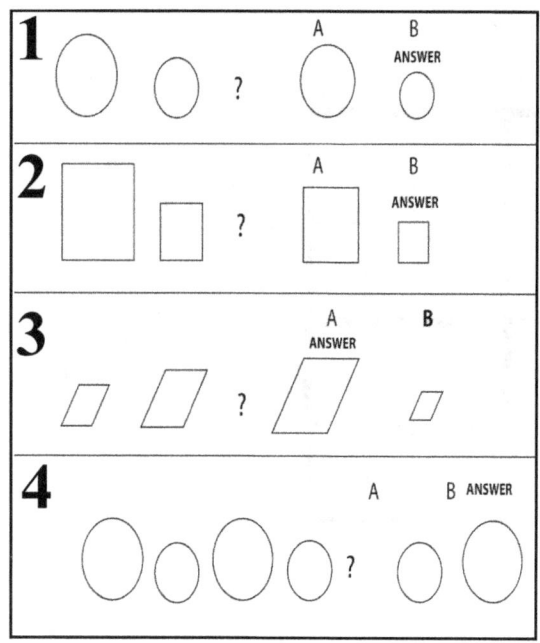
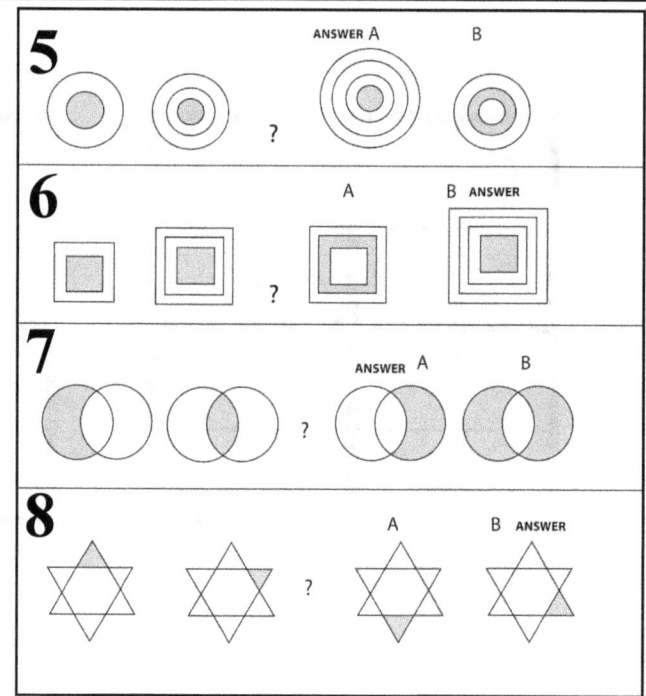
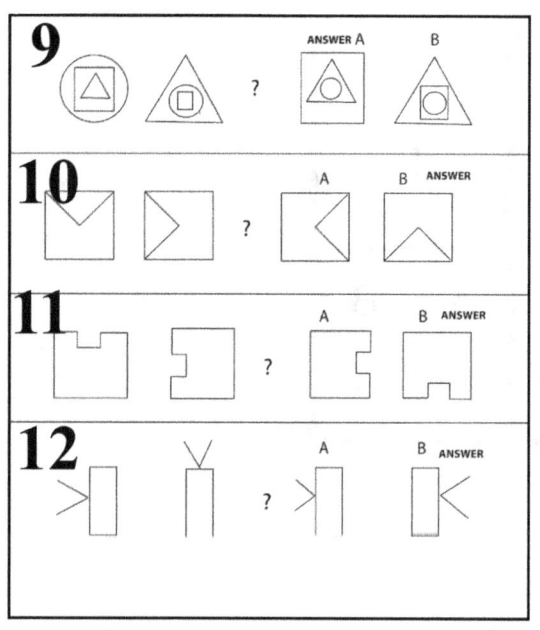
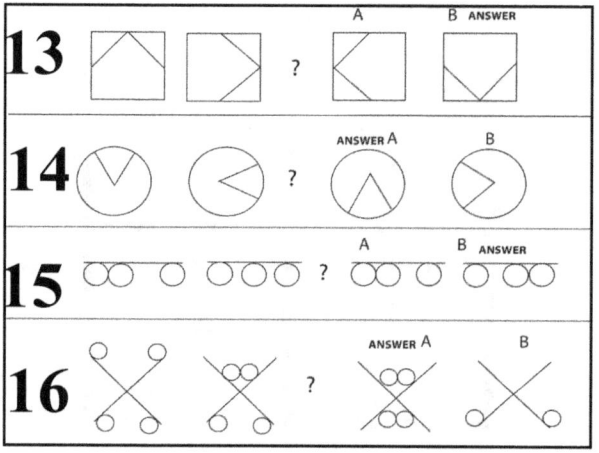

Answers 144

© Gift Of Logic, Inc * Copying prohibited

PICTURE DIFFERENCE

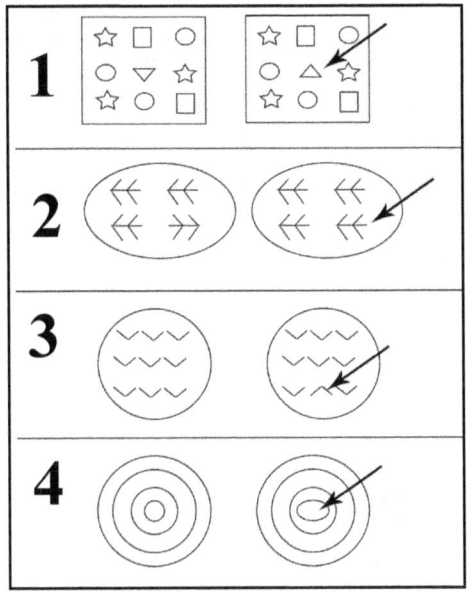
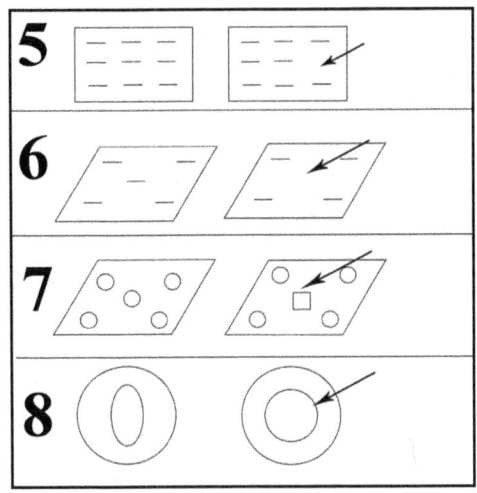
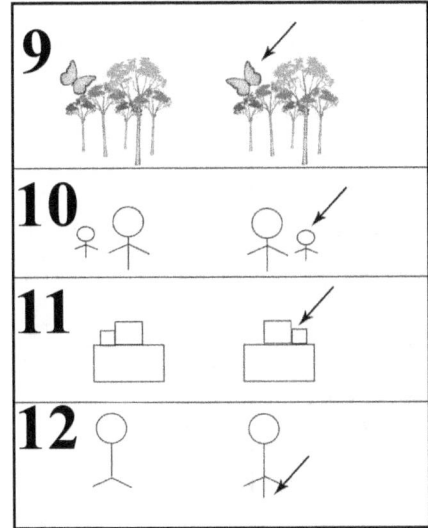
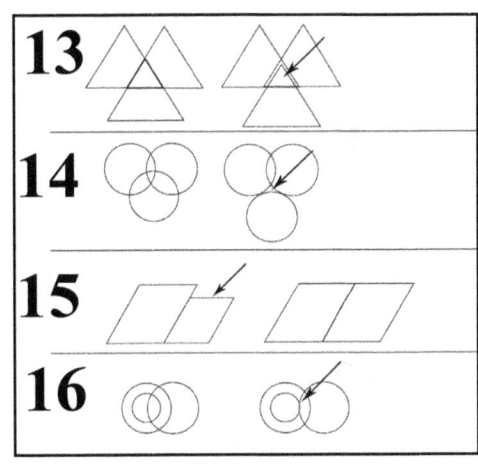

Answers 145
© Gift Of Logic, Inc * Copying prohibited

PICTURE ANALOGY

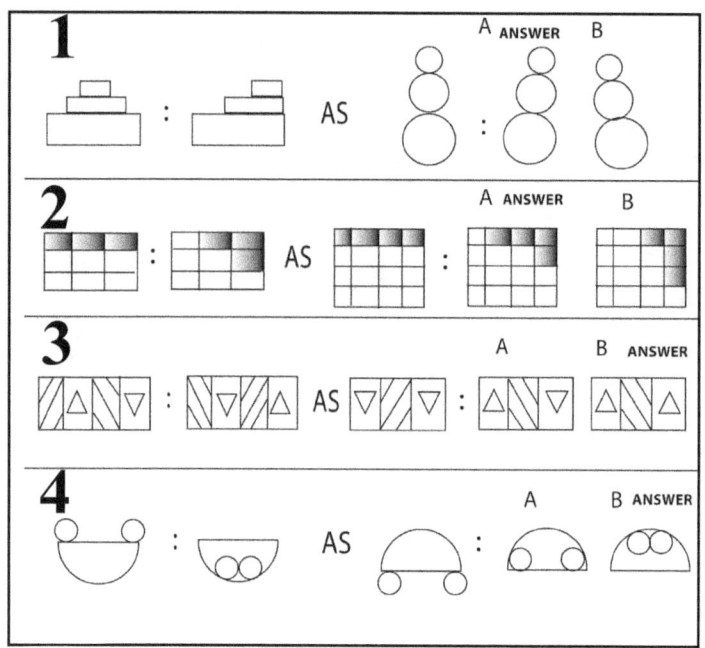

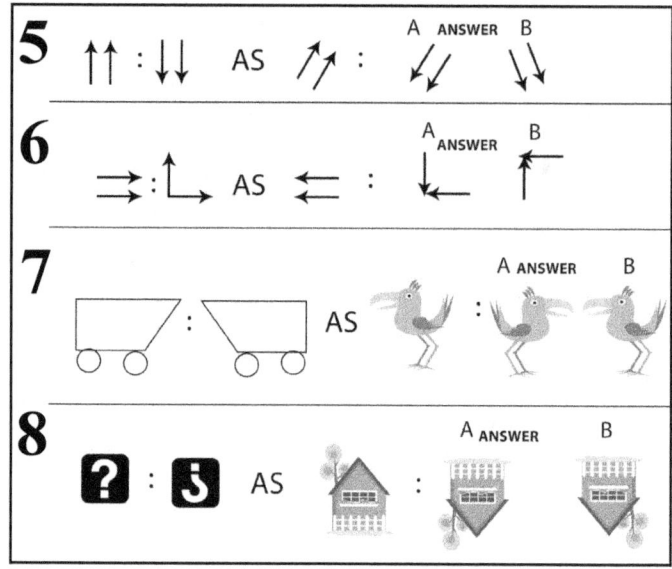

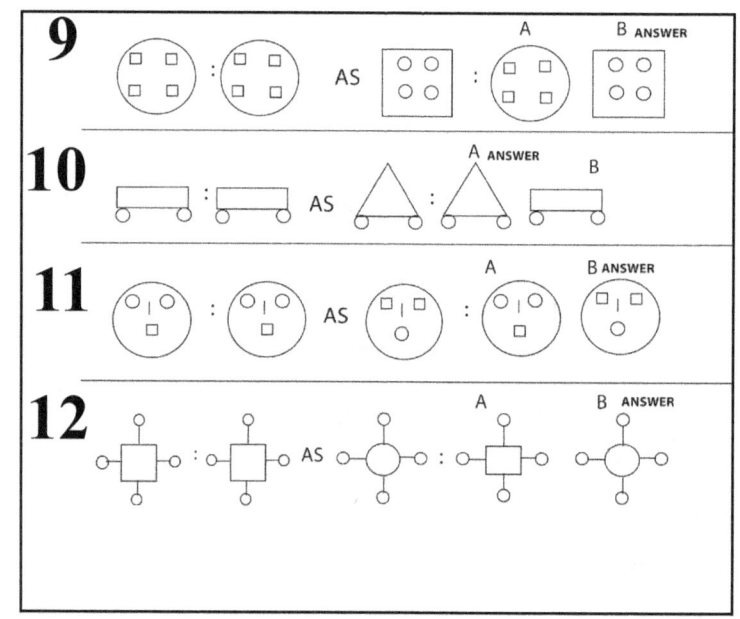

Answers 146
© Gift Of Logic, Inc * Copying prohibited

ODD PICTURE

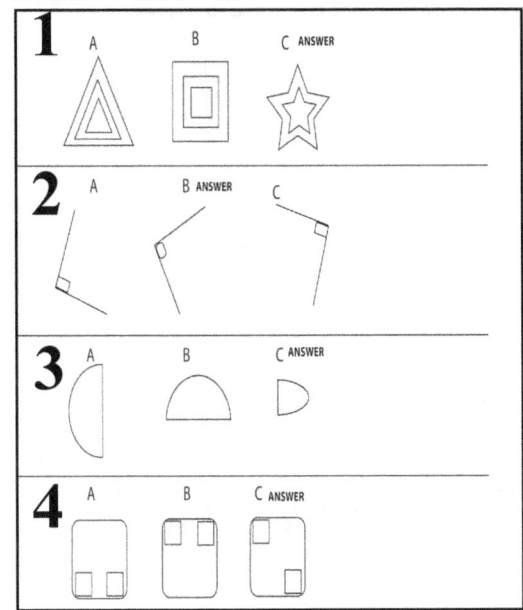
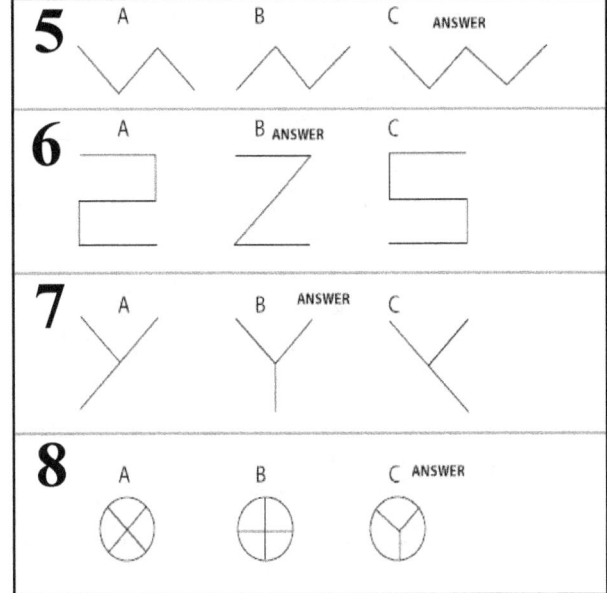
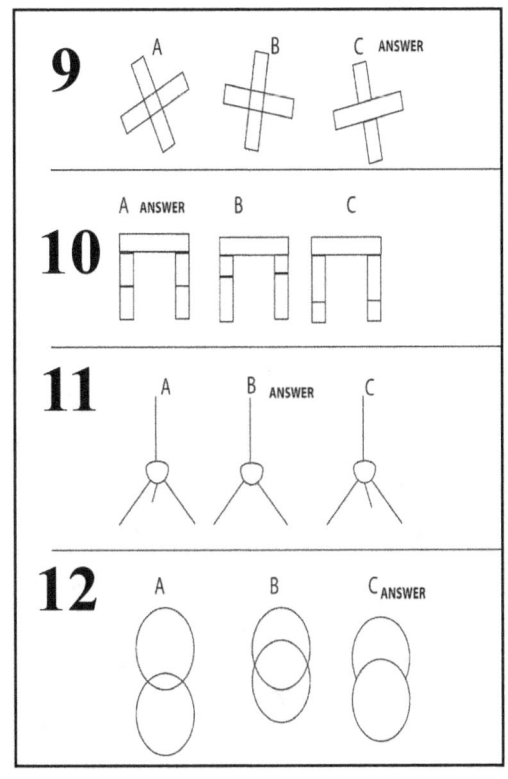
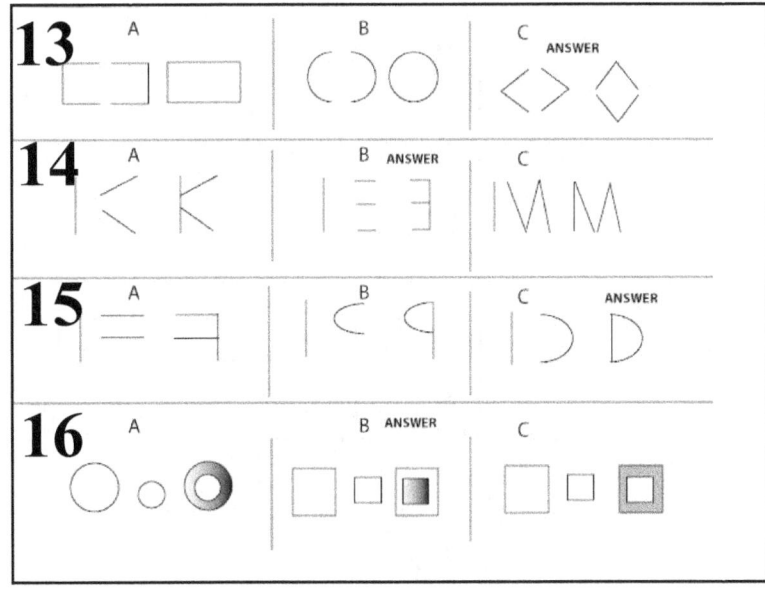

Answers 147
© Gift Of Logic, Inc * Copying prohibited

PATTERN MATCHING

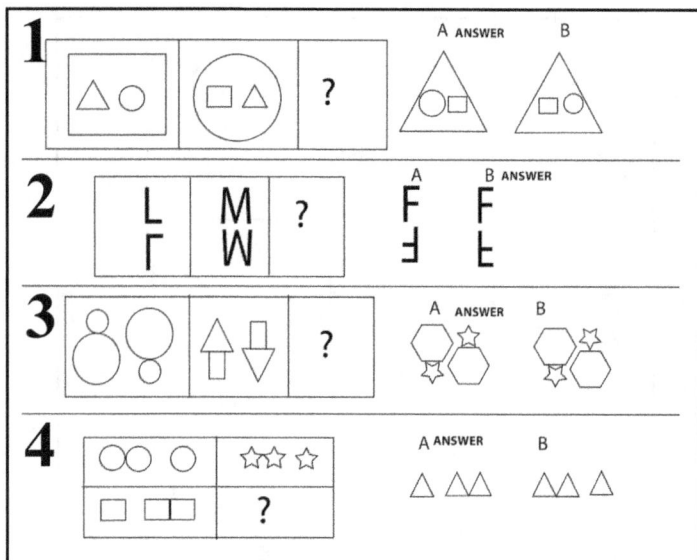
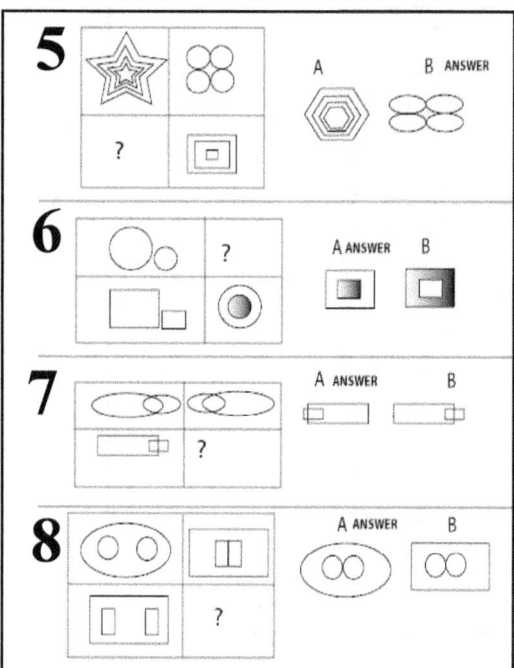
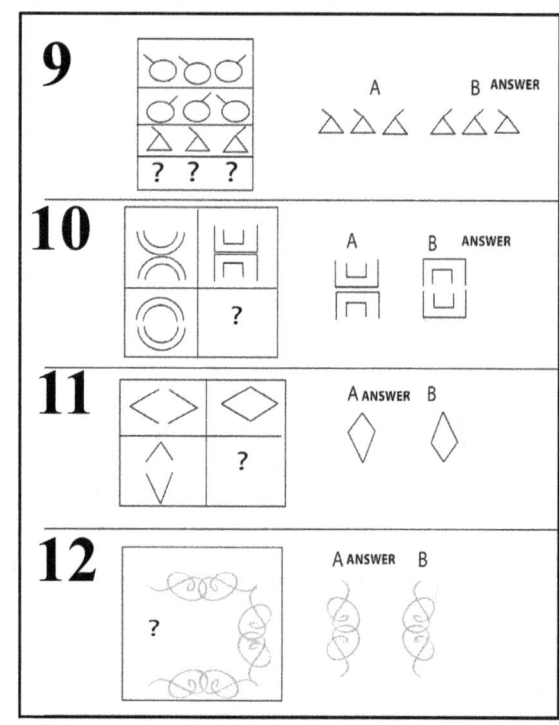

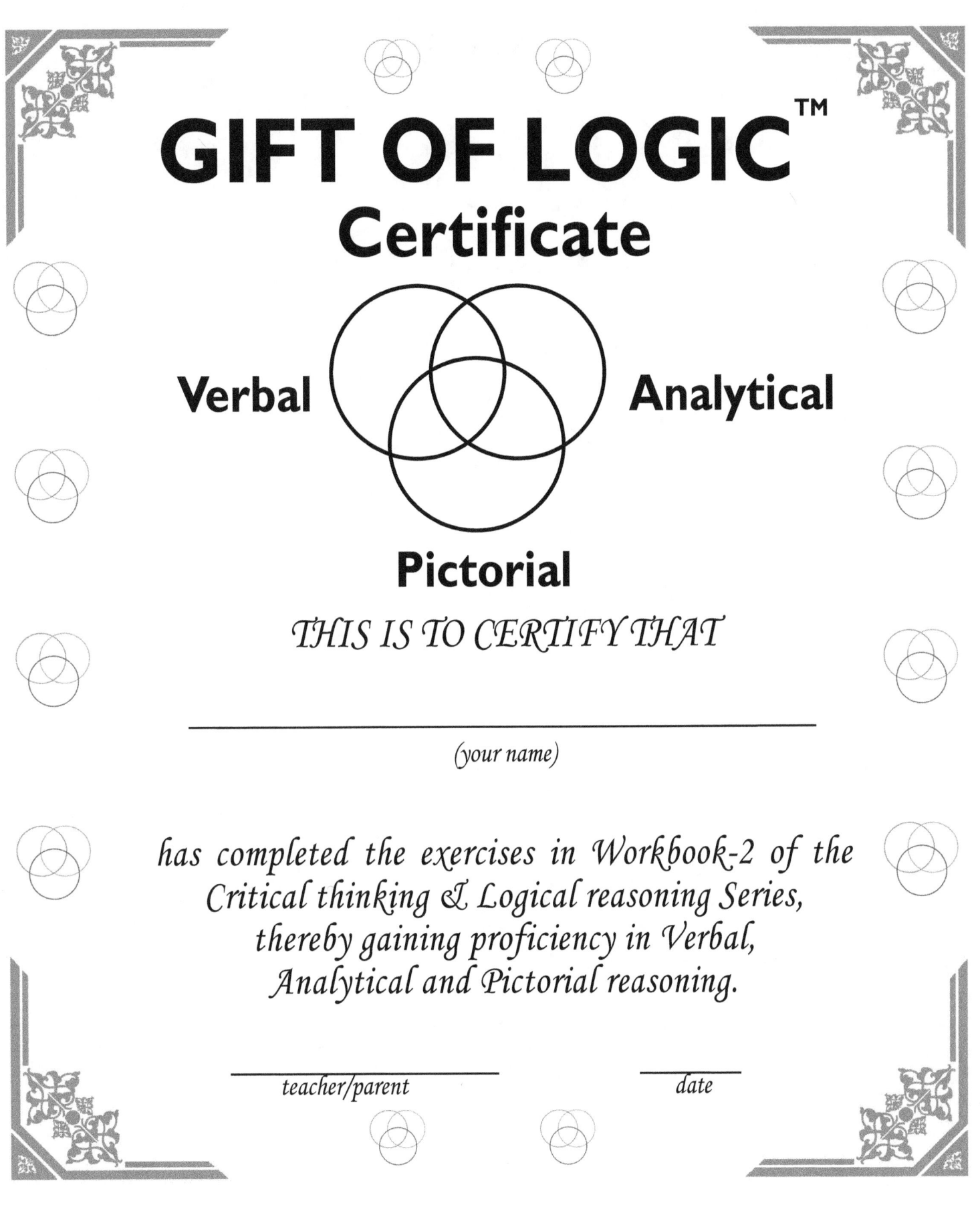

NOTES